Necessary Conditions for an Extremum

PURE AND APPLIED MATHEMATICS

A Series of Monographs

1. KENTARO YANO. Integral Formulas in Riemannian Geometry (1970)
2. S. KOBAYASHI. Hyperbolic Manifolds and Holomorphic Mappings (1970)
3. V. S. VLADIMIROV. Equations of Mathematical Physics (A. Jeffrey, editor; A. Littlewood, translator) (1970)
4. B. N. PSHENICHNYI. Necessary Conditions for an Extremum. (L. Neustadt, translation editor; K. Makowski, translator) (1971)
5. L. NARICI, E. BECKENSTEIN, and G. BACHMAN. Functional Analysis and Valuation Theory (1971)

In Preparation:

W. BOOTHBY and G. L. WEISS (eds.). Geometry and Harmonic Analysis of Symmetric Spaces

Y. MATSUSHIMA. Differentiable Manifolds (E. J. Taft, editor; E. T. Kobayashi, translator)

D. PASSMAN. Infinite Group Rings

L. DORNHOFF. Group Representation Theory

Necessary Conditions for an Extremum

B. N. PSHENICHNYI
Institute of Cybernetics
Kiev, U.S.S.R.

Translation edited by
LUCIEN W. NEUSTADT

Translated by
KAROL MAKOWSKI

Department of Electrical Engineering
University of Southern California
Los Angeles, California

MARCEL DEKKER, INC., New York 1971

This book was originally published in Russian under the title "Neobkhodimye Usloviya Ekstremuma" by Nauka, Moscow, 1969.

MARCEL DEKKER, INC.
95 Madison Avenue, New York, New York 10016

LIBRARY OF CONGRESS CATALOG CARD NUMBER: 76-152570
ISBN NO.: 0-8247-1556-X

PRINTED IN THE UNITED STATES OF AMERICA

FOREWORD TO THE RUSSIAN EDITION

The last ten to twelve years have been a period of extremely rapid development for the theory of extremal problems and for methods of solving such problems. A large number of problems, interesting from a theoretical standpoint and important from a practical standpoint, has attracted the attention of many mathematicians and engineers. And this is not surprising. It is now difficult to name any field of knowledge in which, in one form or another, extremal problems do not arise and in which it is not essential for the development of these fields that such problems be solved. Among such fields are automatic control theory, economics, and even biology. Each of these sciences brings forth its own extremal problems and awaits the answers to two questions: What is the qualitative character of solution, and how does one find a solution?

The first of these questions - what is the qualitative character of a solution - urges mathematicians to look for the most complete necessary conditions for an extremum, since it is precisely such conditions which permit us to foretell the general structure of a solution. The chain-reaction character of the stream of newly arising problems has begun to indicate clearly that we are indeed in need of general conditions for an extremum, i. e., of conditions of a type which may be applied to a broad class of problems. In this way, we can eliminate the necessity of developing a new theory for every concrete case. On the other hand, the particular problems that had already been solved provided a basis of confidence that such conditions could be formulated. Moreover, they showed that such conditions would not be too distant from concrete problems - so that the application of such conditions to a given problem would resemble an occupation of previously prepared positions rather than an assault on a fortress.

Extremum problems are not new in mathematics. They have been encountered and solved during the entire history of mathematics. But an intensive and systematic investigation of such problems has been begun comparatively recently when, on the one hand, the demands of economics and automatic control made the solving of such problems an urgent matter, and, on the other hand, the appearance of electronic computers provided researchers with a powerful tool with whose aid problems could be solved to the point of obtaining a final numerical result. If we do not now speak of the Calculus of Variations and of problems of minimizing functions under equality type constraints, i.e., of problems for which necessary conditions for an extremum were obtained a long time ago, then the beginning of the new stage of development of extremal problem theory can be dated to 1939. In this year, the Soviet mathematician L. V. Kantorovich created methods for solving a new class of problems - linear programming problems. After this, a theory for linear programming was widely developed in the works of G. Dantzig and of many other authors, both abroad and in the U.S.S.R.

The next stage in the development of necessary conditions for an extremum was the elaboration of convex programming theory. A central place in this theory is held by the Kuhn-Tucker theorem, which gives necessary conditions for an extremum, and which was the source for a number of algorithms. The differential form of the Kuhn-Tucker theorem can also be applied to non-convex programming problems in a finite-dimensional space, and makes it possible to formulate necessary conditions for an extremum in such problems.

In elaborating on the necessary conditions for linear and convex programming problems, the principle which lies at the heart of all of the constructions was clarified. This principle was summed up by G. Zoutendijk in his monograph, "Methods of Feasible Directions". The essence of this principle consists in the entirely obvious fact that if, at a given point, there exists a direction which does not lead out of the admissible domain and along which the objective function decreases, then such a point cannot be a mininum point. On the basis of this principle, necessary conditions for an extremum were constructed for a broad class of problems with smooth constraints in a finite-dimensional space.

At the same time that a theory of finite-dimensional extremal problems, i.e., of mathematical programming, was being developed, another class of problems, namely, optimal control problems, was being investigated. The crucial step in this investigation was the formulation of necessary conditions for an extremum in the form of the Pontryagin maximum principle. There is no need to dwell in detail on the importance of this formulation to the entire theory of automatic control, and on the large number of works which it brought forth. Taking into account the direction which we shall take in this book, we should, in the first place, note that the proof of the maximum principle, which is due to V. G. Boltyanskii, was, to some extent, a sensation, since it made use of methods for which it is difficult to find an analog in the previously developed theory of mathematical programming. In this connection, the following question arose: Is it possible to prove the maximum principle with the aid of ideas and methods from the classical calculus of variations and mathematical programming? An affirmative answer to the preceding question would have not only a purely esthetic value, but would also have another, practical side. Namely, it would allow us to apply to optimal control problems the computational methods which have been developed for mathematical programming.

The embedding of optimal control theory into a general theory of necessary conditions was first carried out by A. A. Milyutin and A. Y. Dubovitskii. The great importance of their work lies in the fact that they succeeded in formulating, in a refined form, necessary conditions for an extremum which can be applied to a broad class of problems. Moreover, their work made clear which part of the proof of the maximum principle could be put in a general framework, and which part could be attributed to the specific character of the optimal control problems, i.e., to the presence of the ordinary differential equation constraints. The specific character of the constraints in the form of ordinary differential equations was most fully reflected in the works of R. V. Gamkrelidze, who formulated the ideas of sliding regimes and of quasi-convex sets. Based on the notion of a quasi-convex set, R. V. Gamkrelidze gave a new proof of the Pontryagin maximum principle which clearly distinguished between the general variational problem and the specific character of the differential constraints.

The work in optimal control theory had an extremely rich influence on a general theory of necessary conditions for an extremum. Indeed, this work made possible a revelation of the underlying principles, and also made it possible to develop techniques for constructing necessary conditions and to look upon a broad class of problems from a unified viewpoint. Moreover, a profound and systematic study permitted one to construct necessary conditions for problems containing functions which are not differentiable in the usual sense. In terms of directionally differentiable functions,

H. Halkin and L. W. Neustadt formulated a very general theorem concerning necessary conditions. This theorem can be applied for solving a broad class of problems, including optimal control problems.

It is interesting to note that the proof both of the basic theorems of A. A. Milyutin and A. Y. Dubovitskii and of the theorems of H. Halkin and L. W. Neustadt require tools which have been known in mathematics for a considerable time. And the fact that these results have only been obtained in the last five to six years points up the large amount of work which was carried out with regard to understanding some of the underlying principles and elaborating on some of the basic concepts.

This book is devoted to a presentation of a theory of necessary conditions for an extremum. Our method of presentation is deductive, i.e., first, general results are stated, and then we show how these results can be particularized to specific problems. This manner of presentation seems to be justified at the present time, since there is a great number of works devoted to the derivation of necessary conditions for specific problems, and these works fully paved the way for a development of an abstract presentation. Further, in the theory of necessary conditions, two parts can be earmarked. These parts may, somewhat conditionally, be named as follows: Formal Conditions for an Extremum and Computational Methods.

Formal Conditions for an Extremum are presented in this book in Chapter IV. They consist of a collection of theorems which assert that, if the function being minimized as well as the domain over which the minimization takes place satisfy certain hypotheses, then a certain relation holds at a minimum point. But these theorems do not tell us how to write down concretely, for a given problem, this relation. To construct a useful relation, it is necessary to develop an apparatus of techniques for evaluating certain quantities. In order to make these general statements understandable, we shall explain them for the special case of the problem of finding the minimum of a function of a single variable.

In order that the minimum be achieved at some point, it is necessary that the derivative of the function at this point be zero. In the terminology which we have presented, this is equivalent to a formal condition for an extremum. But if methods for evaluating derivatives of quite complicated functions had not been developed, then it would have been impossible to write down the stated condition in a useful form for any realistic problem. The first three chapters of this book are devoted to an investigation of techniques for evaluating certain quantities. Only after these techniques have been sufficiently developed, are formal conditions for an extremum presented.

Any general theory is of value only to the extent that it permits one to look upon a sufficiently broad class of problems from a unified viewpoint. Therefore, the relatively large fifth chapter is devoted to an illustration of how the constructed theory can be applied to specific problems. Each of the problems which we shall consider is far from trivial, and a quite large number of works has been devoted to investigating them. Some of the problems which we shall investigate, for example, the Chebyshev approximation problem and the moment problem, have by themselves a very general character and numerous applications in Economics and Optimal Control Theory.

In Chapter V, we consider only those optimal control problems for which a result can be obtained without an investigation of the special features introduced by the differential equations constraints. This is because a consideration of these constraints would slightly lead us away from the general direction of the book.

To make the presentation complete, we state, in the introduction, some basic facts from Functional Analysis and from the theory of convex sets which we shall use in our presentation. We have not confined ourselves to finite-dimensional spaces because to refuse to present material which makes use of more general spaces would have resulted in a considerable impoverishment in the material which we could present. Also, it would then have been impossible to treat a number of problems for which, as a matter of fact, the complicated theory had been constructed.

Therefore, the reader who is not too familiar with Functional Analysis may acquaint himself, in the introduction, with all of the facts which are necessary to understand the remainder of the book, all the more since there are very few such facts. The reader who is familiar with the basic results of Functional Analysis can begin his reading with the first chapter.

It is also necessary to make a remark on the method of presenting references to the literature which we adopted in this book. In the main presentation, we make only the most necessary references regarding results which are used, but not proved, in the book.

On the other hand, at the end of the book, in the short bibliography, we present references which indicate in which works certain theorems were proved, what relation some given result has with others, etc.

This book has been written on the basis of a series of lectures which the author gave at the Second All-Union School on Optimization Methods, in the city of Shemakha, July 6-26, 1967. The author is sincerely grateful to the chairman of the Organizing Committee of the School, corresponding member of the Academy of Sciences of the USSR, N.N. Moiseev, for his invitation to read these lectures, and also for numerous fruitful discussions, and for his attention to this work.

I also consider it my pleasant duty to express my acnowledgement to my colleagues in the Institute of Cybernetics of the Academy of Sciences of the Ukrainian SSR, whose help with the work on this book can hardly be overestimated.

B. N. Pshenichnyi

FOREWORD TO THE AMERICAN EDITION

The theory of necessary conditions for an extremum is a field of mathematics for which the greatest contribution was made by Soviet and American scientists. Thus, I am very pleased that a translation of my small book will be published in the USA. I hope that it will be conducive to mutual understanding and friendly contacts between Soviet and American scientists.

I have added to the present edition a small bibliography of works with which I became acquainted after I had written the main text of the book.

B. N. Pshenichnyi

<u>Note to the reader.</u> The symbol ||| will be used to denote the end of a proof.

TABLE OF CONTENTS

FOREWORD TO THE RUSSIAN EDITION v
FOREWORD TO THE AMERICAN EDITION xvi

INTRODUCTION. ELEMENTS OF FUNCTIONAL ANALYSIS AND CONVEX SETS 1
1. Some Basic Concepts of Functional Analysis 1
2. Convex Sets 22
3. Convex Functionals 35

CHAPTER I. PROPERTIES OF CONVEX FUNCTIONALS 39

CHAPTER II. CONVEX PROGRAMMING IN BANACH SPACES 54

CHAPTER III. QUASI-DIFFERENTIABLE FUNCTIONALS 68

CHAPTER IV. NECESSARY CONDITIONS FOR AN EXTREMUM IN GENERAL MATHEMATICAL PROGRAMMING PROBLEMS 82

CHAPTER V. NECESSARY CONDITIONS FOR AN EXTREMUM IN CONCRETE PROBLEMS 120

1. The Classical Mathematical Programming Problem 121
2. Mathematical Programming with a Continuum of Constraints 122
3. Theorems for Minimax Problems 126
4. Chebyshev Approximation Problems 130
5. A Linear Optimal Control Problem with Phase Constraints 142
6. A Duality Principle in Convex Programming 150

7. Systems of Convex Inequalities. Helly's Theorem 157
8. The Moment Problem 166
9. A Discrete Maximum Principle 185

SHORT BIBLIOGRAPHY 201

LITERATURE 210

NOTES AND SUPPLEMENTARY BIBLIOGRAPHY TO AMERICAN EDITION 217

SUPPLEMENTARY LITERATURE 223

SUBJECT INDEX 226

INTRODUCTION

ELEMENTS OF FUNCTIONAL ANALYSIS AND CONVEX SETS

Functional analysis is the mathematical apparatus on which the construction of the theory of necessary conditions for minimization problems is based. As a matter of fact, only a few basic concepts and a few theorems are used to construct the theory. These concepts are, first of all, the ideas of weak convergence, compactness, and a separation theorem for convex sets. For the reader's convenience, we shall briefly, and without proofs, state those basic facts of Functional Analysis which are necessary for an understanding of the subsequent material. We shall take them from [1]. Incidentally, the majority of the theorems stated in the sequel – with the exception of a few basic ones – are immediate consequences of the definitions, and may be proved by the reader himself, if he wishes to test whether he correctly understands the introduced definitions.

1. Some basic concepts of Functional Analysis

Definition 1. A family τ of subsets of a set X forms a _topology_ on this set if τ contains the empty set $\emptyset$, the set X itself, any union (of any number) and any intersection of a finite number of its sets. The pair (X, τ) is said to be a _topological space_. The sets of τ are said to be _open_ sets. Any open set which contains a point p is said to be a _neighborhood of_ p. _A neighborhood of a set_ A is any open set which contains A. A point $p \in A$ is said to be an _interior point_ of A if there exists a neighborhood of p entirely contained in A.

The following Lemma is an obvious consequence of Definition 1.

Lemma 1. A set in a topological space is open if and only if it contains a neighborhood of each of its points.

Definition 2. A family β of subsets of X is said to be a _basis of the topology_ τ if every set of β is contained in the family τ, and if every set in τ is the union of sets from the family β.

In order that a family β be a basis of some topology, it is necessary and sufficient that, for every pair of sets U, $V \in \beta$, and every $x \in U \cap V$, there exists a set $W \in \beta$ such that $x \in W \subset U \cap V$ and, furthermore, that the union of all sets from β coincides with X.

If a basis β is given, then the topology τ consists of all sets which are arbitrary unions of sets in β. For example, the usual topology on an n-dimensional space can be given by means of the family β which consists of the sets defined by the inequalities

$$\sum_{i=1}^{n} (x_i - y_i)^2 < R^2,$$

where $y = (y_1, \ldots, y_n)$ is any vector, and R is any positive number.

Definition 3. A set $Y \subset X$ is said to be closed if its complement $X \setminus Y$ is open. The intersection of all closed sets which contain a set A is said to be the closure of A, and is denoted by $\overline{A}$.

Having defined a topology τ on some set X, the notion of convergence of sequences can be introduced.

Definition 4. A sequence x_n, $n \to \infty$, converges to a point x_o, where $x_n \in X$ and $x_o \in X$, if, for any neighborhood of x_o, there exists a number N such that all points x_n, $n \geq N$, belong to that neighborhood.

In the sequel, we shall consider only so-called Hausdorff topological spaces, i.e., topological spaces (X, τ) in which any two distinct points have nonintersecting neighborhoods. In such spaces, sets consisting of single points are evidently closed, and every convergent sequence has a unique limit point, so that we can write

$$\lim_{n \to \infty} x_n = x_o .$$

The notion of convergence introduced above for the case of an n-dimensional space is entirely equivalent to the usual concept of convergence which is given in mathematical analysis courses [44]. Namely, a sequence of vectors x^k converges to a vector x^o if, for any $\varepsilon > 0$, there exists an N such that, for $k \geq N$,

$$\sum_{i=1}^{n} (x_i^k - x_i^o)^2 < \varepsilon^2 .$$

Let there now be given two topological spaces (X_1, τ_1) and (X_2, τ_2). It is said that a mapping $f: X_1 \to X_2$ (from X_1 into X_2) is given if a unique element $y \in X_2$ is associated with every $x \in X_1$. This element y is denoted by f(x). If $U \subset X_1$, then the set of all points f(x), as x ranges over U, is denoted by f(U). If $V \subset X_2$, then $f^{-1}(V)$ is the set of all points x such that $f(x) \in V$. The set f(U) is said to be the image of U, and $f^{-1}(V)$ is said to be the pre-image of V.

Definition 5. A mapping $f: X_1 \to X_2$ is said to be

continuous at a point x_o if, for any open set $V \subset X_2$ with $f(x_o) \in V$, there exists a neighborhood U of x_o such that $f(U) \subset V$. A mapping $f: X_1 \to X_2$ is said to be continuous if it is continuous at every point $x \in X_1$.

A mapping f is continuous if and only if, for any open set $V \subset X_2$, the pre-image $f^{-1}(V)$ is also an open set. Indeed, let V be an open set in X_2, and let $x \in f^{-1}(V)$. By definition of continuity at x, and because $f(x) \in V$, there is a neighborhood U_x such that $f(U_x) \subset V$. Then it is obvious that

$$f^{-1}(V) = \bigcup_{x \in f^{-1}(V)} U_x .$$

But the set in the right-hand side of this equation, as the union of open sets, is open.

If the more familiar definition of continuity, by means of sequences, is considered, then it can easily be proved that, if f(x) is continuous at a point x_o, then $x_n \to x_o$ implies that $f(x_n) \to f(x_o)$.

In exactly the same manner, if A is a set which is closed in the sense of Definition 3, then $x_n \to x_o$ and $x_n \in A$ implies that $x_o \in A$.

There is an important class of spaces – metric spaces – for which the concepts of continuity of functions and of closed sets which we have introduced are equivalent to these concepts when they are stated in the language of sequences. In such spaces, a function is continuous at a point x_o if and only if $x_n \to x_o$ implies that $f(x_n) \to f(x_o)$, and a set A is closed if and only if $x_n \to x_o$ and $x_n \in A$ implies that $x_o \in A$.

Definition 6. A set X is said to be a metric space if there is a real-valued function $\rho(x, y)$, defined on the set of all pairs $x, y \in X$, with the following properties:

1) $\rho(x, y) \geq 0$ and $\rho(x, y) = 0$ only if $x=y$,

2) $\rho(x, y) = \rho(y, x)$,

3) $\rho(x, y) \leq \rho(x, y) + \rho(y, z)$.

In such a space, a topology may be introduced by means of the basis consisting of all sets of the form

$$S(x_o, r) = \{x: x \in X, \ \rho(x, x_o) < r\}.$$

In particular, the usual space of n-dimensional vectors is a metric space. A metric in this space may be given by the formula

$$\rho(x, y) = \left(\sum_{i=1}^{n} (x_i - y_i)^2 \right)^{1/2}$$

The concept of compactness and the related concept of sequential compactness play a very important role in the sequel.

<u>Definition</u> 7. A <u>cover</u> of a set A in a topological space X is a family of open sets whose union contains A. A set A is said to be compact if, from any cover of A, it is possible to extract a finite subcover. A set A is said to be sequentially compact if, from any sequence $\{x_n\}$, $x_n \in A$, a convergent sequence can be chosen such that its limit point belongs to A.

Let us indicate some properties of compact sets.

<u>Lemma</u> 2. a) <u>If the mapping</u> f: $X_1 \to X_2$ <u>is continuous, then the image</u> f(U) <u>of a compact set</u> U <u>is compact.</u>

b) <u>A compact set in a Hausdorff space is closed.</u>

c) <u>In a metric space, the concepts of sequential compactness and compactness coincide.</u>

By virtue of a well-known theorem of Weierstrass in analysis, a set in an n-dimensional space is sequentially compact if and only if it is closed and bounded.

Let X_1 and X_2 be two topological spaces with topologies τ_1 and τ_2. In order to be able to deal with functions of the form $f(x_1, x_2)$, where $x_1 \in X_1$ and $x_2 \in X_2$, we shall introduce the concept of the <u>direct product of two spaces</u> X_1 and X_2, which we shall denote by $X_1 \times X_2$. Namely, $X_1 \times X_2$ consists of all pairs (x_1, x_2), where $x_1 \in X_1$ and $x_2 \in X_2$, with a topology defined by means of the following basis: The basis of $X_1 \times X_2$ consists of all sets of the form $\{(x_1, x_2): x_1 \in U_1, x_2 \in U_2, U_1 \in \tau_1, U_2 \in \tau_2\}$. Such a set will be denoted by $U_1 \times U_2$. In accordance with this definition, if a topology on the real line (which is a one-dimensional space) is defined by means of the basis which consists of all open intervals, $\{x: a < x < b\}$, then the basis in n-dimensional space consists of all sets of the form $\{x: x = (x_1, \ldots, x_n), -r_i < x_i - x_i^o < r_i \text{ for } i=1, \ldots, n\}$ where the r_i are given positive numbers, and $(x_1^o, \ldots, x_n^o)$ is a given point in the n-dimensional space.

Thus, if an n-dimensional space is considered to be the direct product of n one-dimensional spaces, the basis in the product space is given by all possible parallelepipeds.

It is now clear what we should understand by continuity of a function $f: X_1 \times X_2 \to Y$, i.e., of a function $f(x_1, x_2)$ defined for $x_1 \in X_1$ and $x_2 \in X_2$, and taking on its values in Y. Namely, such a function is continuous if it is continuous when considered as a function from the topological space $X_1 \times X_2$ into the topological space Y.

In what follows, we shall deal with topological spaces which possess the structure of a linear space.

Definition 8. A space L is said to be *linear* if, for any $x, y \in L$, the operations of addition $x + y \in L$ and multiplication by a real number $\lambda x \in L$ are defined and satisfy the conditions

1. $(x+y)+z = x+(y+z)$.

2. $x+y = y+x$.

3. There exists an element $\theta \in L$ such that, for all $x \in L$, $x+\theta = x$ and $0x = \theta$.

4. $(\lambda+\mu)x = \lambda x + \mu x$.

5. $\lambda(x+y) = \lambda x + \lambda y$.

6. $(\lambda\mu)x = \lambda(\mu x)$.

7. $1 \cdot x = x$.

The element θ plays the role of "zero" in the space and will often be denoted by a "zero", if this does not lead to any misunderstanding.

As an example of a linear space, we indicate the space E^n of n-dimensional vectors, in which the operations of addition of vectors and multiplication by real numbers are defined in the usual way.

A topology may be given on a linear space L. In such a case, the space L becomes a <u>linear topological</u> space if the topology is such that the functions $g(x, y) = x+y$ and $f(\lambda, x) = \lambda x$, which map $L \times L \to L$ and $E^1 \times L \to L$, are continuous. Here, E^1 is the space of all real numbers with its topology given with the aid of the basis consisting of all sets of the form

$$S(\lambda_o, r) = \{\lambda: |\lambda - \lambda_o| < r\}.$$

A linear space is said to be normed if, with each element x in the space, there is associated a real number $\|x\|$ satisfying the following conditions:

1) $\|x\| \geq 0$ and $\|x\| = 0$ if and only if $x = 0$,

2) $\|\lambda x\| = |\lambda| \cdot \|x\|$,

3) $\|x + y\| \leq \|x\| + \|y\|$.

Such a number will be called the norm of an element x. Given a norm on a linear space, it is possible to define a topology with a basis consisting of all sets of the form

$$S(x_o, r) = \{x: \|x - x_o\| < r\}. \tag{1}$$

In such a topology, a sequence x_n converges to x_o if and only if

$$\|x_n - x_o\| \underset{n \to \infty}{\to} 0.$$

A normed space is a metric space, since a metric can be defined on it by the formula

$$\rho(x, y) = \|x - y\|.$$

By virtue of the remark which was previously made, in a normed space the concepts of sequential compactness and compactness coincide, and the concepts of a continuous function and a closed set can be stated in the language of sequences.

For example, a set A is closed if and only if $x_n \to x_o$, $x_n \in A$, implies that $x_o \in A$.

A function $\mu(x)$, defined on a normed space, which takes on its values in the space E^1 (i. e., $\mu(x)$ is a real number) is continuous at a point x_o if and only if, for every $\varepsilon > 0$, there exists $\delta > 0$ such that

$$|\mu(x) - \mu(x_o)| < \varepsilon$$

whenever $\|x - x_o\| < \delta$. If we take into account the facts that one basis of the topology on E^1 is given by all intervals of the form $\{\lambda: |\lambda - \lambda_o| < \varepsilon\}$, and that a basis for the topology on a normed space is given by Formula (1), then we can easily see that the given definition of continuity agrees completely with Definition 5.

A function $\mu(x)$ whose values belong to the set of real numbers is normally called a functional.

A function A(x) from B_1 into B_2, where B_1 and B_2 are normed spaces, is said to be an operator.

In the sequel, a normed space will be denoted by the letter B. In the study of a normed space B, a special role is played by the space which is conjugate to B. This is the space of all continuous linear functionals defined on B, which is denoted by B^*. Let us recall that a functional is said to be linear if

$$\mu(x+y) = \mu(x) + \mu(y)$$

and

$$\mu(\lambda x) = \lambda\mu(x).$$

The elements of the space B^* will be denoted by letters with an asterisk, e.g., x^*, y^*, etc.

The space B^* is a linear space since the operations of addition of functionals and multiplication by numbers can be defined as follows:

$$(x^* + y^*)(x) = x^*(x) + y^*(x),$$

$$(\lambda x^*)(x) = \lambda x^*(x).$$

It is obvious that if x^* and $y^* \in B^*$, then the functionals $x^* + y^*$ and λx^* are also linear and continuous.

It is equally easy to verify that if $B = B_1 \times B_2$, then $B^* = B_1^* \times B_2^*$. Thus, if $x^* \in B^*$, then

$$x^*(x) = x_1^*(x_1) + x_2^*(x_2)$$

for some $x_1^* \in B_1^*$ and $x_2^* \in B_2^*$, for any $x = \{x_1, x_2\}$.

A topology on the space B^* can be defined in two different ways. The first such topology, the so-called strong topology, is given with the aid of the basis consisting of all sets of the form

$$V(X, r, x_o^*) = \{x^*: \sup_{x \in X} |x^*(x) - x_o^*(x)| < r\},$$

where X is any <u>bounded</u> subset of B, i.e., a set such that $\|x\| \leq R$ for some $R < \infty$ and all $x \in X$. In this topology, B^* turns out to be itself a normed space, if the norm of a functional is defined by the formula

$$\|x^*\| = \sup_{\|x\| \leq 1} x^*(x),$$

and if we take as a basis for the topology the family of all sets of the form

$$S^*(x_o^*, r) = \{x^*: \|x^* - x_o^*\| < r\}.$$

It is relevant to notice that the following dual formula holds

$$\|x\| = \max_{x^* \in S^*} x^*(x),$$

where

$$S^* = \{x^*: \|x^*\| \leq 1\}.$$

The fact that the bases $V(X, r, x_o^*)$ and $S^*(x_o^*, r)$ generate the same topology follows from the fact that $S^*(x_o^*, r)$ coincides with $V(X, r, x_o^*)$, if

$$X = S = \{x: \|x\| \leq 1\},$$

and it is easy to verify that a set $V(X, r, x_o^*)$ can be made up as the union of sets of the form $S^*(x_o^*, r)$.

The second topology, the <u>weak* topology</u> on the space

B^*, is defined with the aid of the basis consisting of all sets of the form

$$W(X, r, x_o^*) = \{x^*: |x^*(x) - x_o^*(x)| < r \text{ for } x \in X\},$$

where X is a subset of B which contains a finite number of elements $x_i \in B$, $i=1, \ldots, k$.

In accordance with the existence of these two topologies, each term, such as sequential compactness, compactness, closure, and convergence has two meanings, depending on whether the strong or the weak* topology is under consideration. For example, convergence can be either strong or weak*.

We note that the family of open sets which forms the strong topology is larger than the family of open sets which forms the weak* topology. This may be verified immediately from the definitions of $V(X, r, x_o^*)$ and $W(X, r, x_o^*)$. Therefore (see the definition of convergence in topological spaces), strong convergence of a sequence of functionals x_n^* to x_o^* implies weak* convergence. However, the converse is not true, since x_n^* converges strongly to x_o^* if

$$\|x_n^* - x_o^*\| \underset{n\to\infty}{\longrightarrow} 0,$$

but x_n^* converges to x_o^* in the weak* topology, if, for any $x \in B$,

$$x_n^*(x) \underset{n\to\infty}{\longrightarrow} x_o^*(x).$$

In exactly the same way, a set $X^* \subset B^*$ is compact in the strong topology only if it is compact in the weak* topology, but the converse is again not true.

Things are different with regard to the closure concept. If a set is closed in the strong topology, it can turn out not to be closed in the weak* topology. Indeed, a closed set is the complement of an open set. The family of open sets in the strong topology is larger than the family of open sets in the weak* topology. Because of this, if the complement of some set is open in the strong topology, it can turn out not to be open in the weak* topology. On the other hand, if a set is weak* closed, then it is also strongly closed.

Let us now consider a special kind of functional defined on the space B^*. Let $x_o \in B$. Then $x^*(x_o)$ is a linear functional on B^*, since, with each x^* we associate the number $x^*(x_o)$, and, moreover,

$$(x^* + y^*)(x_o) = x^*(x_o) + y^*(x_o)$$

$$(\lambda x^*)(x_o) = \lambda x^*(x_o).$$

Let us show that the functional so defined is continuous in both topologies. To do this, we must show that, for any

x_o^*, the set of all x^* such that

$$|x^*(x_o) - x_o^*(x_o)| < \varepsilon$$

is open, i.e., this set is a neighborhood of x_o^*. But this set coincides with the set

$$V(x_o, \varepsilon, x_o^*) = W(x_o, \varepsilon, x_o^*),$$

which, by definition, is open in both topologies.

The concept of convergence in a normed space is closely related to the concept of a fundamental sequence.

A sequence is said to be fundamental if, for any $\varepsilon > 0$, there exists a number N such that

$$\|x_n - x_m\| < \varepsilon$$

whenever $n, m > N$. It is easy to show that if $x_n \to x_o$, then the sequence x_n, $n=1, \dots,$ is fundamental. Indeed, since

$$\|x_n - x_o\| \underset{n\to\infty}{\to} 0,$$

then for a given ε, there exists a number N such that

$$\|x_n - x_o\| < \frac{\varepsilon}{2}$$

for all $n>N$. Now let $m>N$. Then

$$\|x_n - x_m\| = \|(x_n - x_o) + (x_o - x_m)\| \leq \|x_n - x_o\| + \|x_o - x_m\| < \varepsilon.$$

Here, the third property of a norm which was indicated in its definition has been used.

Let us now consider a normed space B which has the property that, for every fundamental sequence x_n, there exists an element $x_o \in B$ such that $x_n \rightarrow x_o$. Such a space is said to be <u>complete</u>. A complete normed space is called a <u>Banach space</u>.

The following theorem holds [1].

<u>Theorem 1. If B is a Banach space, then a subset X^* of B^* is weak* compact if and only if it is weak* closed and if there exists a constant R such that</u>

$$\|x^*\| \leq R \text{ for all } x^* \in X^*.$$

In every linear space, the <u>sum of two subsets</u> X_1 and X_2 can be defined as follows: $X_1 + X_2$ is the set of all elements $x_1 + x_2$, where $x_1 \in X_1$, $x_2 \in X_2$. Let the space under consideration be B^*, and let X_1^* and X_2^* be subsets thereof.

<u>Theorem 2. If X_1^* and X_2^* are weak* closed, and if X_2^* is weak* compact, then $X_1^* + X_2^*$ is weak* closed.</u>

In a linear space, <u>multiplication of a set by a scalar</u> can also be defined. If $X \subset L$, then λX is the set consisting of all elements λx, $x \in X$. These two operations already define a sum of the form $\lambda_1 X_1 + \lambda_2 X_2$, consisting, in an obvious way, of all elements $\lambda_1 x_1 + \lambda_2 x_2$, where $x_1 \in X_1$ and $x_2 \in X_2$.

We shall present two examples of Banach spaces which one most often comes across in extremal problems.

The space E^n. This is the space of all n-dimensional vectors $x = (x_1, \ldots, x_n)$. Addition and scalar multiplication in this space are defined in the usual way:

$$x + y = (x_1 + y_1, \ldots, x_n + y_n),$$

$$\lambda x = (\lambda x_1, \ldots, \lambda x_n).$$

The norm in E^n is given by the formula

$$\|x\| = \sqrt{\sum_{i=1}^{n} x_i^2}.$$

Every linear functional in $(E^n)^*$ is given by the formula

$$x^*(x) = \sum_{i=1}^{n} a_i x_i,$$

where $a = (a_1, \ldots, a_n)$ is also a vector in n-dimensional space. In this way, to any $x^* \in (E^n)^*$ there corresponds a unique vector a. Conversely, the preceding formula shows that, to every $a \in E^n$ there corresponds a continuous functional x^*. Moreover, as is easily verified, the sum of two functionals corresponds to the sum of the corresponding vectors, and the product of a functional and a scalar corresponds to the product of the vector and the scalar. For this reason, the space $(E^n)^*$ can be identified with an n-dimensional vector space.

The strong topology in $(E^n)^*$ is defined by means of the norm

$$\|x^*\| = \sup_{\|x\| \le 1} \sum_{i=1}^{n} a_i x_i = \sqrt{\sum_{i=1}^{n} a_i^2} = \|a\|.$$

Thus, $(E^n)^*$ not only is a space of n-dimensional vectors, but also the norm of a functional turns out to coincide with the norm of the corresponding vector. This means that the strong topology in $(E^n)^*$ coincides with the topology of E^n, and $(E^n)^* = E^n$.

It is easy to show that, in a finite-dimensional space, the weak* topology and the strong topology coincide, because they define the same families of open sets. For this reason, in the space that is conjugate to E^n, there is no difference between strong and weak* convergence and closure, and the concept of sequential compactness coincides with the concept of compactness.

When dealing with a finite-dimensional space, it is convenient to introduce the idea of a <u>scalar product</u> of one vector with another. This product is denoted by (a, x):

$$(a, x) = \sum_{i=1}^{n} a_i x_i .$$

In this notation,

$$x^*(x) = (a, x),$$
$$\|x\| = \sqrt{(x, x)}.$$

The space $C[0,1]$. This is the space of all continuous functions defined on the interval $[0,1]$, with norm given by the formula

$$\|x\| = \max_{0 \le t \le 1} |x(t)|.$$

In this space, convergence of a sequence of functions $x_n(t)$ to a function $x_o(t)$ coincides with uniform convergence. The space of continuous functionals C^* consists of all functionals of the form

$$x^*(x(t)) = \int_0^1 x(t)\,dg(t), \qquad g(0) = 0,$$

where $g(t)$ is a function of bounded variation which is continuous from the right at all t with $0<t<1$. The norm $\|x^*\|$ is given by the formula

$$\|x^*\| = \operatorname*{Var}_{0 \le t \le 1} g(t),$$

where

$$\operatorname*{Var}_{0 \le t \le 1} g(t) = \sup \sum_{i=1}^{k} |g(t_i) - g(t_{i-1})|,$$

and the supremum is taken with respect to all subdivisions t_i, $i=0, \ldots, k$, $t_o = 0$, $t_k = 1$.

In concluding this brief presentation of the elements of Functional Analysis, let us introduce the idea of a differential of an operator. Let an operator $A(x)$ map a space B into B_1.

We say that an operator $A(x)$ is <u>Fréchet differentiable</u> at a point $x_o \in B$ if there exists a <u>linear</u> operator $A'(x_o): B \to B_1$ such that

$$A(x) - A(x_o) = A'(x_o)(x - x_o) + r(x_o, x - x_o),$$

where the function $r(x_o, z)$ is such that

$$\lim_{z \to 0} \frac{\|r(x_o, z)\|}{\|z\|} = 0.$$

An operator $A(x)$ is <u>Gâteaux differentiable</u> if

$$A(x_o + \lambda e) - A(x_o) = \lambda A'(x_o)\, e + r(x_o, \lambda e),$$

where the function $r(x_o, \lambda e)$ is such that

$$\lim_{\lambda \to 0} \frac{\|r(x_o, \lambda e)\|}{\lambda} = 0.$$

The operator $A'(x_o)$ in the first case is said to be a <u>Fréchet differential</u>, and, in the second case, a <u>Gâteaux differential</u>. If $A'(x_o)$ is a Fréchet differential, then it is also a Gâteaux differential, but the converse is not always true.

We remind the reader that an operator A is said to be <u>linear</u> if it is continuous and if

$$A(x + y) = A(x) + A(y),$$

$$A(\lambda x) = \lambda A(x).$$

For any linear operator it is possible to define an <u>adjoint</u>

operator, which we shall denote by A^*. If A is a linear operator, then, for any $x_1^* \in B_1^*$, the functional μ on B, defined by

$$\mu(x) = x_1^*(Ax), \quad x \in B,$$

is, of course, linear and continuous. Thus, with every $x_1^* \in B_1^*$, we can associate a functional $x^* \in B^*$ defined by

$$x^*(x) = x_1^*(Ax) \text{ for all } x \in B.$$

Let us denote this functional x^* by $A^* x_1^*$. In this way, we may consider A^* to be an operator from B_1^* into B^*. A^*, which turns out to be linear, will be called the adjoint of A. Therefore, the adjoint operator is defined by the following relations:

$$A^*: B_1^* \to B^*,$$

$$(A^* x_1^*)(x) = x_1^*(Ax).$$

2. Convex sets. The concept of a convex set plays a central role in the study of extremal problems. In essence, the entire theory of necessary conditions is an extended corollary to separation theorems for convex sets.

Definition 9. A set X in a linear space L is said to be convex if $x_1 \in X$ and $x_2 \in X$ imply that

$$\lambda_1 x_1 + \lambda_2 x_2 \in X$$

for all

$$\lambda_1, \lambda_2 \geq 0, \quad \lambda_1 + \lambda_2 = 1.$$

In other words, a set is convex if and only if it contains the entire line segment joining any two of its points.

The following property follows at once from the definition of convexity. If a set X is convex, then, whenever it contains the points x_i, $i=1, \ldots, n$, it also contains the point

$$x = \sum_{i=1}^{n} \lambda_i x_i, \tag{2}$$

where

$$\sum_{i=1}^{n} \lambda_i = 1, \quad \lambda_i \geq 0.$$

The proof of this fact can be carried out by induction. For $n = 2$, it follows from the definition. Suppose that it is true for $n = k-1$. Let $n = k$ in (2), and suppose that $\lambda_i > 0$ for each $i=1, \ldots, k$. If some $\lambda_i = 0$, then the problem reduces to the already considered case of $n = k-1$. Let

$$\lambda_i' = \frac{\lambda_i}{\lambda}, \quad i=2, \ldots, k, \quad \lambda = \lambda_2 + \ldots + \lambda_k.$$

Then

$$\sum_{i=2}^{k} \lambda_i' = 1,$$

and the point

$$\bar{x} = \sum_{i=2}^{k} \lambda_i' x_i = \frac{1}{\lambda} \sum_{i=2}^{k} \lambda_i x_i \in X$$

by the induction hypothesis. Then

$$x = \lambda_1 x_1 + \lambda\bar{x}, \quad x_1 \in X, \quad \bar{x} \in X,$$

and, since $\lambda_1 \geq 0$, $\lambda \geq 0$ and

$$\lambda_1 + \lambda = \sum_{i=1}^{k} \lambda_i = 1,$$

$x \in X$ by the definition of a convex set, as was to be proved.

Now let B be a Banach space. A basic property of convex sets, a property that makes these sets such a valuable instrument for the investigation of extremal problems, is given by the following separation theorem.

Theorem 3. *Any two non-intersecting convex sets X and Y in a Banach space, one of which contains an interior point, can be separated by a non-zero linear functional. This means that there exists a linear functional* $x^* \in B^*$, *not identically equal to zero, such that*

$$x^*(x) \leq x^*(y)$$

for all $x \in X$ and $y \in Y$.

Remark 1. In a Banach space, a point x_0 is an interior point of a set X if, for some $r > 0$,

$$\{x: \|x - x_0\| < r\} \subset X.$$

Remark 2. If a space is finite-dimensional, then the

requirement in Theorem 3 that one of the sets must have an interior point can be omitted.

If the sets X and Y are closed, then Theorem 3 can be sharpened.

Theorem 4. *If X and Y are non-intersecting, closed, convex sets in B, and if Y is compact, then there exist a continuous linear functional $x^* \in B^*$ and constants c and $\varepsilon > 0$ such that*

$$x^*(x) \leq c - \varepsilon < c \leq x^*(y)$$

for all $x \in X$ and $y \in Y$.

Corollary. *If X is a closed, convex set in B, and $x_0 \notin X$, then there exists a functional $x_0^* \in B^*$ such that*

$$x_0^*(x) \leq x_0^*(x_0) - \varepsilon, \quad \text{for all } x \in X,$$

for some $\varepsilon > 0$.

Indeed, a set which consists of only one point is obviously compact |||.

We shall have to consider convex sets in the space B^*, conjugate to a Banach space B. In this space, so-called *regularly convex* sets are particularly important.

Definition 10. A set $X^* \subset B^*$ is said to be *regularly convex* if, for every functional $x_0^* \notin X^*$, there exists an element

$x_o \in B$ such that

$$x^*(x_o) < x_o^*(x_o) - \varepsilon$$

for all $x^* \in X^*$ and some $\varepsilon > 0$.

Definition 10 specifies regularly convex sets in terms of separability. The following theorem [52] shows how such sets can be characterized in terms of convexity and weak* closedness.

Theorem 5. *A set X^* is regularly convex if and only if it is convex and weak* closed.*

Thus, if a set X^* is convex and weak* closed, then we can separate the set and every functional x_o^* which does not belong to this set by means of some element $x_o \in B$.

If a set X is not convex, it is possible to construct the smallest convex set which contains it. Such a set is said to be the *convex hull* of X and is denoted by co X or [X]. The given definition of a convex hull is valid for a set X in any linear space. Moreover, if the linear space L under consideration is also a linear topological space, then it is possible to introduce the concept of the *closed convex hull*, $\overline{\text{co}}\, X$, which is defined as the smallest closed, convex set containing X.

It is easy to see that co X consists of all points x

which can be represented as a convex combination of points from X, i. e., of all points x of the form

$$x = \sum_{i=1}^{n} \lambda_i x_i, \quad x_i \in X, \quad \sum_{i=1}^{n} \lambda_i = 1, \quad \lambda_i \geq 0.$$

This follows immediately from the fact that the set of all points which are convex combinations of points of X is convex and, as has been shown previously, that every convex set must contain every convex combination of any of its points.

Let us present some properties of convex hulls.

Lemma 3. For arbitrary subsets X and Y of a linear space L:

1) $\mathrm{co}\,(\alpha X) = \alpha\,\mathrm{co}\,X$, $\mathrm{co}\,(X+Y) = \mathrm{co}\,X + \mathrm{co}\,Y$.

If L is a linear topological space, then

2) $\overline{\mathrm{co}}\,(X) = \overline{\mathrm{co}\,(X)}$.

3) If $\overline{\mathrm{co}}\,X$ and $\overline{\mathrm{co}}\,Y$ are compact, then $\overline{\mathrm{co}}\,(X \cup Y) = \mathrm{co}\,(\overline{\mathrm{co}}\,X \cup \overline{\mathrm{co}}\,Y)$.

In particular, if X and Y are compact and convex, then $\overline{\mathrm{co}}\,(X \cup Y) = \mathrm{co}\,(X \cup Y)$.

In a finite-dimensional space, the convex hull of a set has one particular property which turns out to be extremely useful, and which gives rise to a series of subtle results in the theory of Chebyshev approximations and in the theory of moments.

Theorem 6. *If X is a subset of an n-dimensional space* E^n, *then any point in* co X *can be represented as a convex combination of not more than* n+1 *points of* X.

Proof. As we have indicated, any point in co X can be represented in the form

$$x = \sum_{i=1}^{k} \lambda_i x_i, \quad x_i \in X.$$

Suppose that $k > n+1$ and that every $\lambda_i > 0$. (If some $\lambda_i = 0$, then the number k can be decreased.)

Let us introduce the vectors y_i in (n+1)-dimensional space formed as follows:

$$y_i = \begin{pmatrix} x_i \\ 1 \end{pmatrix}.$$

Since $k > n+1$, the number of vectors y_i is greater than the dimension of the space in which they lie. Thus, all these vectors are linearly dependent, i. e., there exist numbers α_i, not all zero, such that

$$\sum_{i=1}^{k} \alpha_i y_i = 0,$$

or, in a "by-component" form,

$$\sum_{i=1}^{k} \alpha_i x_i = 0, \quad \sum_{i=1}^{k} \alpha_i = 0.$$

Since not all of the α_i are zero, and their sum is zero, at least one of the α_i is positive. Now let

$$\lambda_i(\varepsilon) = \lambda_i - \varepsilon\alpha_i.$$

Then

$$\sum_{i=1}^{k} \lambda_i(\varepsilon) = \sum_{i=1}^{k} \lambda_i - \varepsilon \sum_{i=1}^{k} \alpha_i = 1.$$

Moreover, since $\lambda_i > 0$, for small ε, $\lambda_i(\varepsilon) > 0$.

Let us now increase ε from zero until one of the $\lambda_i(\varepsilon)$ equals zero. Since at least one of the α_i is positive, this must take place for some $\varepsilon = \varepsilon_o$, $\varepsilon_o > 0$. Then

$$\lambda_i' = \lambda_i(\varepsilon_o) \geq 0,$$

and at least one of the λ_i' equals zero.

Now,

$$x = \sum_{i=1}^{k} \lambda_i' x_i.$$

Indeed,

$$\sum_{i=1}^{k} \lambda_i' x_i = \sum_{i=1}^{k} (\lambda_i - \varepsilon_o \alpha_i) x_i = \sum_{i=1}^{k} \lambda_i x_i - \varepsilon_o \sum_{i=1}^{k} \alpha_i x_i = x.$$

Thus, x has been represented in the form of a convex combination of k points $x_i \in X$, and at least one coefficient λ_i'

vanishes. This means that, if x can be represented as a convex combination of k points $x_i \in X$ and $k > n+1$, then this point can be also represented in the form of a convex combination of k-1 points.

Essentially, the theorem follows from this statement, since by means of the described procedure, the number k can be decreased until it equals n+1. |||

Let us now consider a special class of convex sets – convex cones.

Definition 11. A set K in a linear space is said to be a convex cone if it is convex and if $x \in K$ implies that $\lambda x \in K$ for every $\lambda > 0$.

It is easy to verify that K is a convex cone if and only if $x, y \in K$ imply that $x+y \in K$ and $\lambda x \in K$ for every $\lambda > 0$.

If the original space B is a Banach space, then every convex cone $K \subset B$ induces some other cone K^* in B^*, which is called the dual cone or the conjugate cone to the cone K.

By definition,

$$K^* = \{x^*: x^* \in B^*,\ x^*(x) \geq 0 \text{ for all } x \in K\}.$$

Since $x^* \in K^*$, $y^* \in K^*$, and $\lambda > 0$ imply that

$$x^*(x) + y^*(x) \geq 0, \quad \lambda x^*(x) \geq 0$$

for all $x \in K$, K^* is also a convex cone.

Let us present some properties of the cone K^*. It is clear that the zero functional $x^* = 0$ always belongs to K^*. If $\overline{K} \neq B$, i.e., if $\overline{K}$ does not coincide with the entire space, then K^* contains elements other than zero. Indeed, if $\overline{K} \neq B$, then there exists an element x_o which does not belong to the closed, convex set $\overline{K}$. Then, by the corollary to Theorem 4, there exists a functional x_o^* such that

$$x_o^*(x) \leq x_o^*(x_o) - \varepsilon$$

for all $x \in K$, or, if we set $x_1^* = -x_o^*$, then

$$x_1^*(x) \geq x_1^*(x_o) + \varepsilon .$$

Let us show that

$$x_1^*(x) \geq 0$$

for all $x \in K$. Indeed, if $x_1^*(x_1) < 0$ for some $x_1 \in K$, then, since $\lambda x_1 \in K$ for all $\lambda > 0$,

$$x_1^*(\lambda x_1) = \lambda x_1^*(x_1) \to -\infty$$

as $\lambda \to \infty$. At the same time, since $\lambda x_1 \in K$, it follows from the construction of x_1^* that

$$x_1^*(\lambda x_1) \geq x_1^*(x_o) + \varepsilon .$$

The contradiction which has just been obtained shows that

$$x_1^*(x) \geq 0$$

for all $x \in K$, and x_1^* is a non-zero functional belonging to the cone K^*

Lemma 4. 1) $(\overline{K})^* = K^*$.

2) $x \in \overline{K}$ *if and only if*

$$x^*(x) \geq 0 \text{ for all } x^* \in K^*.$$

3) *If* x *is an interior point of* K, *then* $x^*(x) > 0$ *for all non-zero* $x^* \in K$.

Proof. 1) Since B is a normed space, the closure $\overline{K}$ of a cone K consists of the points in K and of the points x_o such that there exists a sequence $x_n \in K$ with $x_n \to x_o$. Thus, if $x^* \in K^*$,

$$x^*(x_n) \geq 0 \text{ for all } n,$$

and, by the continuity of x^*,

$$x^*(x_o) \geq 0$$

for any $x_o \in \overline{K}$. It follows from this that $x^* \in (\overline{K})^*$. Thus, $K^* \subset (\overline{K})^*$. On the other hand, since $\overline{K} \supset K$, by definition of a dual cone, $K^* \supset (\overline{K})^*$. Therefore, $K^* = (\overline{K})^*$.

2) If $x \in \overline{K}$, then by the definition of $(\overline{K})^*$,

$$x^*(x) \geq 0 \text{ for all } x^* \in (\overline{K})^*.$$

But we have just shown that $K^* = (\overline{K})^*$. Therefore, the desired inequality holds if $x \in \bar{K}$. Now let x_o be such that

$$x^*(x_o) \geq 0$$

for all $x^* \in K^*$. Suppose that $x_o \notin \bar{K}$. Then, as was shown in proving that K^* contains non-zero elements if $\bar{K} \neq B$, there exists a functional $x_1^* \in K^*$ such that

$$x_1^*(x) \geq x_1^*(x_o) + \varepsilon$$

for some $\varepsilon > 0$ and for all $x \in \bar{K}$. But, since $\lambda x \in K$ for any $x \in K$ and $\lambda > 0$, then, by letting λ tend to zero, we obtain that $0 \in \bar{K}$. Substituting 0 into the preceding inequality, we obtain

$$-\varepsilon \geq x_1^*(x_o) .$$

This contradicts the fact that $x_1^* \in K^*$ and that $x^*(x_o) > 0$ for all $x^* \in K^*$.

3) Let x_o be an interior point of K; i.e., suppose that, for some $r > 0$,

$$\{x: \|x - x_o\| < r\} \subset K. \tag{3}$$

For any non-zero $x^* \in K^*$,

$$x^*(x) \geq 0$$

for all $x \in K$, and, in particular, for all x satisfying (3).

By definition of the norm of a functional,

$$\|x^*\| = \sup_{\|x\| \leq 1} x^*(x) ,$$

there exists an element e with $\|e\| \leq 1$ such that

$$\|x^*\| \geq x^*(e) \geq \frac{1}{2} \|x^*\|.$$

Let us consider the point

$$x_1 = x_o - \frac{r}{2} e.$$

This point satisfies (3) because

$$\|x_1 - x_o\| = \|\frac{r}{2} e\| = \frac{r}{2} \|e\| < r.$$

Therefore, $x_1 \in K$ and

$$x^*(x_1) = x^*(x_o) - \frac{r}{2} x^*(e) \geq 0,$$

or

$$x^*(x_o) \geq \frac{r}{2} x^*(e) \geq \frac{r}{4} \|x^*\| > 0,$$

as was to be proved. |||

Lemma 5. If K is a convex cone, then the cone K^* is weak* closed.

Proof. By definition of a closed set, it is necessary to show that the complement of K^* is open. For this purpose, it is sufficient, by Lemma 1 of the introduction, to show that, if $x_o^* \notin K^*$, then there is a neighborhood of x_o^* in the weak* topology of B^* which does not have any points in common with K^*.

Let $x_o^* \notin K^*$. Then, by definition of K^*, there exists an $x_o \in K$ such that

$$x_o^*(x_o) = \alpha < 0.$$

Let

$$M = \{x^* : 2\alpha < x^*(x_o) < 0\}.$$

It is obvious that $x_o^* \in M$ and that M does not have any points in common with K^*. We shall show that M is open. Indeed,

$$M = W(x_o, |\alpha|, x_o^*) = \{x^* : |x^*(x_o) - x_o^*(x_o)| < |\alpha|\},$$

which, by definition, is open in the weak* topology of B^*. |||

3. Convex functionals. We shall make a more detailed study of convex functionals in the next chapter. Here, we shall dwell only on the definition of a convex functional and on a few of its simplest properties.

Definition 12. A functional $\mu(x)$, defined for all x in a linear space L, is said to be convex if, for any $\lambda_1, \lambda_2 \geq 0$ with $\lambda_1 + \lambda_2 = 1$, and for any $x_1, x_2 \in L$, the following inequality holds:

$$\mu(\lambda_1 x_1 + \lambda_2 x_2) \leq \lambda_1 \mu(x_1) + \lambda_2 \mu(x_2).$$

If we set $\lambda = \lambda_1$, $1-\lambda = \lambda_2$, then the last inequality can be rewritten in a slightly different, but often used, form

$$\lambda\mu(x_1) + (1-\lambda)\mu(x_2) \geq \mu(\lambda x_1 + (1-\lambda)x_2)$$

for all λ with $0 \leq \lambda \leq 1$.

Let $\mu(x)$ be a convex functional. For fixed $x \in L$ and $e \in L$, let us denote

$$\varphi(t) = \mu(x + te).$$

The function $\varphi(t)$ is a convex function of a one-dimensional variable t, where $-\infty < t < \infty$.

Indeed,

$$\varphi(\lambda_1 t_1 + \lambda_2 t_2) = \mu(x + (\lambda_1 t_1 + \lambda_2 t_2)e) = \mu(\lambda_1(x + t_1 e)$$

$$+ \lambda_2(x + t_2 e)) \leq \lambda_1 \mu(x + t_1 e) + \lambda_2 \mu(x + t_2 e) = \lambda_1 \varphi(t_1) + \lambda_2 \varphi(t_2).$$

Thus, $\varphi(t)$ satisfies the following inequality:

$$\varphi(\lambda_1 t_1 + \lambda_2 t_2) \leq \lambda_1 \varphi(t_1) + \lambda_2 \varphi(t_2). \qquad (5)$$

Let $t_o < t_1 < t_2$, and let

$$\lambda_1 = \frac{t_1 - t_o}{t_2 - t_o}, \quad \lambda_2 = 1 - \frac{t_1 - t_o}{t_2 - t_o}.$$

Note that

$$\lambda_1 t_2 + \lambda_2 t_o = \frac{t_1 - t_o}{t_2 - t_o} t_2 + \quad 1 - \frac{t_1 - t_o}{t_2 - t_o} \quad t_o = t_1.$$

Therefore, Inequality (5) can be rewritten in the form

$$\varphi(t_1) \leq \frac{t_1 - t_o}{t_2 - t_o} \varphi(t_2) + \quad 1 - \frac{t_1 - t_o}{t_2 - t_o} \quad \varphi(t_o),$$

or

$$\frac{\varphi(t_1) - \varphi(t_o)}{t_1 - t_o} \leq \frac{\varphi(t_2) - \varphi(t_o)}{t_2 - t_o}. \qquad (6)$$

Further, let $t_{-1} < t_o < t$. Then

$$\varphi(\lambda_1 t + \lambda_2 t_{-1}) \leq \lambda_1 \varphi(t) + \lambda_2 \varphi(t_{-1}). \tag{7}$$

We set

$$\lambda_1 = \frac{t_o - t_{-1}}{t - t_{-1}}, \quad \lambda_2 = \frac{t - t_o}{t - t_{-1}}.$$

Then it is easy to see that $\lambda_1, \lambda_2 > 0$, $\lambda_1 + \lambda_2 = 1$ and that

$$\lambda_1 t + \lambda_2 t_{-1} = t_o.$$

Now Inequality (7) can be rewritten in the form

$$\varphi(t_o) \leq \frac{t_o - t_{-1}}{t - t_{-1}} \varphi(t) + \frac{t - t_o}{t - t_{-1}} \varphi(t_{-1})$$

or

$$(t_o - t_{-1})\varphi(t_o) + (t - t_o)\varphi(t_o) \leq (t_o - t_{-1})\varphi(t) + (t - t_o)\varphi(t_{-1}),$$

$$\frac{\varphi(t_o) - \varphi(t_{-1})}{t_o - t_{-1}} \leq \frac{\varphi(t) - \varphi(t_o)}{t - t_o}. \tag{8}$$

The inequalities (6) and (8) which we have just proved allow us to draw a conclusion which we shall state in the form of a lemma.

Lemma 6. *If $\varphi(t)$ is a convex function of the scalar variable t, then the function*

$$\gamma(t) = \frac{\varphi(t) - \varphi(t_o)}{t - t_o},$$

<u>defined for</u> $t > t_o$, <u>is non-decreasing as</u> t <u>increases, and is bounded from below.</u>

<u>Corollary.</u> <u>Let</u>

$$\varphi(t) = \mu(x + te),$$

<u>where</u> $\mu(x)$ <u>is a convex function.</u> <u>Let</u> $t_o = 0$ <u>and</u> $t = \lambda$. <u>Then the quotient</u>

$$\gamma(\lambda) = \frac{\mu(x + \lambda e) - \mu(x)}{\lambda},$$

<u>defined for</u> $\lambda > 0$, <u>is non-decreasing as</u> λ <u>increases, and is bounded from below.</u>

<u>Lemma</u> 7. <u>If</u> $\mu(x)$ <u>is a convex functional, then, at every point</u> x <u>and for every direction</u> e, <u>the directional differential</u>

$$\frac{\partial \mu(x)}{\partial e} = \lim_{\lambda \to +0} \frac{\mu(x + \lambda e) - \mu(x)}{\lambda}.$$

<u>exists.</u>

<u>Proof.</u> Since the quotient

$$\frac{\mu(x + \lambda e) - \mu(x)}{\lambda}$$

is decreasing as $\lambda \to +0$ and is bounded from below, it follows from a well-known theorem in Mathematical Analysis that the quotient tends to a limit as $\lambda \to +0$. But the existence of this limit implies the existence of the directional differential. |||

CHAPTER I

PROPERTIES OF CONVEX FUNCTIONALS

Let $\mu(x)$ be an arbitrary convex functional. Throughout the sequel, we shall suppose that if $X \subset B$ and if

$$\|x\| \leq k$$

for all $x \in X$, then there exists a number C such that $\mu(x) \leq C$ for all $x \in X$. We shall call such convex functionals <u>bounded</u> functionals.

Let us introduce the following definition.

<u>Definition</u> 1.1. The set $M(x_o) \subset B^*$ defined by

$$M(x_o) = \{x^* : x^* \in B^*,\ \mu(x) - \mu(x_o) \geq x^*(x - x_o) \text{ for all } x \in B\},$$

is said to be <u>the set of support functionals to</u> $\mu(x)$ <u>at</u> x_o. The existence of such a set for any $x_o \in B$ is a consequence of the following theorem.

<u>Theorem</u> 1.1. $M(x_o)$ <u>is non-empty, convex, weak* closed and bounded.</u>

<u>Proof.</u> Let $B_o = E^1 \times B$, and, in this space, let us

consider the set

$$Z = \{(\alpha, x) : \alpha > \mu(x)\}.$$

This set is convex because the relations $\alpha_1 > \mu(x_1)$, $\alpha_2 > \mu(x_2)$, $\lambda_1 \geq 0$, $\lambda_2 \geq 0$, and $\lambda_1 + \lambda_2 = 1$ imply, by the convexity of $\mu(x)$, that

$$\lambda_1\alpha_1 + \lambda_2\alpha_2 > \lambda_1\mu(x_1) + \lambda_2\mu(x_2) \geq \mu(\lambda_1 x_1 + \lambda_2 x_2).$$

Moreover, Z contains interior points. Indeed, by assumption, there exists a c_o such that

$$\mu(x) \leq c_o$$

for all x which satisfy the inequality $\|x - x_o\| \leq 1$. Let $\alpha_o = c_o + 1$. It is obvious that $(\alpha_o, x_o) \in Z$. Thus, all points (α, x) which satisfy the conditions

$$|\alpha - \alpha_o| \leq \frac{1}{2}, \quad \|x - x_o\| \leq 1, \qquad (1.1)$$

also belong to Z, because, for such (α, x),

$$\alpha \geq \alpha_o - \frac{1}{2} = c_o + \frac{1}{2} > \mu(x)$$

by definition of c_o.

By definition, Z does not have any points in common with the ray

$$L = \{(\alpha, x_o) : \alpha < \mu(x_o)\}.$$

Therefore, there exist a number c and a functional $y^* \in B^*$

[1], not both zero, such that

$$c\alpha + y^*(x) \geq c\mu(x_o) + y^*(x_o) \tag{1.2}$$

for all $(\alpha, x) \in Z$. Setting here $x = x_o$, we obtain

$$c(\alpha - \mu(x_o)) \geq 0$$

for all $\alpha \geq \mu(x_o)$. This implies that $c \geq 0$. But if $c = 0$, then it follows from (1.2) that

$$y^*(x - x_o) \geq 0$$

for all $x \in B$ — and this is impossible, since, if $c = 0$, then $y^* \neq 0$.

Thus, $c > 0$. Setting $\alpha = \mu(x)$ in (1.2), we obtain

$$\mu(x) - \mu(x_o) \geq -\frac{1}{c}\, y^*(x - x_o),$$

i.e., $x^* = \frac{-1}{c} y^*$ belongs to $M(x_o)$. This proves that $M(x_o)$ is non-empty. The fact that $M(x_o)$ is convex and weak* closed can be verified in an elementary manner.

Let us show that $M(x_o)$ is bounded. Suppose the contrary. Then there exists a sequence $\{x_n^*\} \subset M(x_o)$ such that $\|x_n^*\| \to \infty$. For each n, let $y_n \in S$ (where S is the unit ball in B) be such that

$$x_n^*(y_n) + \varepsilon \geq \|x_n^*\| \geq x_n^*(y_n).$$

Setting $x_n = x_o + y_n$, we obtain

$$\mu(x_n) - \mu(x_o) \geq x_n^*(y_n) \geq \|x_n^*\| - \varepsilon,$$

which implies that $\mu(x_n) \to \infty$ as $n \to \infty$. But $\|x_n - x_o\| \leq 1$, contradicting the fact that $\mu(x)$ is bounded on every bounded set. |||

The following theorem makes clear the role which $M(x_o)$ plays in the investigation of extremal problems involving convex functionals.

Theorem 1.2. The directional differential

$$\frac{\partial \mu}{\partial e} = \lim_{\lambda \to +0} \frac{\mu(x_o + \lambda e) - \mu(x_o)}{\lambda}$$

exists for all x_o and e, and is given by the formula

$$\frac{\partial \mu}{\partial e} = \max_{x^* \in M(x_o)} x^*(e). \qquad (1.3)$$

Proof. The existence of $\frac{\partial \mu}{\partial e}$ follows from the results which we obtained in the introduction. It was also proved in the introduction that

$$\varphi(\lambda) = \frac{\mu(x_o + \lambda e) - \mu(x_o)}{\lambda}$$

is a non-decreasing function of λ.

Let us prove that formula (1.3) holds. By definition, for any $x^* \in M(x_o)$, we have

$$\mu(x_o + \lambda e) - \mu(x_o) \geq \lambda x^*(e),$$

i. e.,

$$\frac{\mu(x_o + \lambda e) - \mu(x_o)}{\lambda} \geq x^*(e)$$

for all $x^* \in M(x_o)$ and $\lambda > 0$. Therefore,

$$\frac{\partial\mu}{\partial e} \geq \max_{x^* \in M(x_o)} x^*(e).$$

Suppose that, for some e_o,

$$\frac{\partial\mu}{\partial e_o} > \max_{x^* \in M(x_o)} x^*(e_o). \tag{1.4}$$

Let us consider the following ray L in $B_o = E^1 \times B$:

$$L = \left\{(\alpha, x) : \alpha = \mu(x_o) + \lambda \frac{\partial\mu}{\partial e_o},\quad x = x_o + \lambda e_o,\ \lambda > 0\right\}.$$

Since $\varphi(\lambda)$ is non-decreasing for increasing λ, $\frac{\partial\mu}{\partial e_o} \leq \varphi(\lambda)$,

and

$$\mu(x_o + \lambda e_o) \geq \mu(x_o) + \lambda \frac{\partial\mu}{\partial e_o}. \tag{1.5}$$

This implies that the previously defined set Z and the ray L do not have any points in common. Indeed, suppose the contrary. This means that, for some $\lambda > 0$, the point (α_1, x_1) where

$$\alpha_1 = \mu(x_o) + \lambda_o \frac{\partial\mu}{\partial e_o},$$

$$x_1 = x_o + \lambda_o e_o,$$

belongs to Z and is an interior point of this set, i. e.,

$$\alpha_1 > \mu(x_1),$$

and, for sufficiently small $\delta > 0$,

$$\alpha_1 - \delta \geq \mu(x_1).$$

But, by definition of α_1 and x_1, the last inequality implies that

$$\mu(x_o) + \lambda_o \frac{\partial\mu}{\partial e_o} - \delta \geq \mu(x_o + \lambda_o e_o),$$

which contradicts (1.5).

Therefore, Z and L do not have any points in common, and thus (Theorem 3 of the introduction) there exist a number c and a functional $y^* \in B^*$ such that

$$c\alpha + y^*(x) \geq c\left(\mu(x_o) + \lambda \frac{\partial\mu}{\partial e_o}\right) + y^*(x_o + \lambda e_o) \qquad (1.6)$$

for all α, x and λ such that $\alpha > \mu(x)$ and $\lambda \geq 0$. Moreover, c and y^* cannot both be zero. Similarly, as was done in the proof of Theorem 1. 1, we can show that $c > 0$. Setting $\alpha = \mu(x)$ and $\lambda = 0$ in (1. 6), we obtain

$$\mu(x) - \mu(x_o) \geq -\frac{1}{c} y^*(x - x_o),$$

i. e., $x^* = -\frac{1}{c} y^* \in M(x_o)$. Next, (1. 6) implies that the following inequality holds:

$$\mu(x) - \mu(x_o) \geq x^*(x - x_o) + \lambda\left[\frac{\partial\mu}{\partial e_o} - x^*(e_o)\right].$$

Setting here $x = x_o$, and taking into account that $\lambda > 0$, we obtain

$$\frac{\partial\mu}{\partial e_o} \leq x^*(e_o),$$

which contradicts (1.4). |||

Remark 1. Formula (1.3) is a generalization of the well-known formula

$$\frac{\partial\mu}{\partial e} = x^*(e),$$

which is satisfied in case the functional $\mu(x)$ is Fréchet differentiable [7] at x_o.

For convex functionals, the set $M(x_o)$ plays the same role that normal derivatives play in the finite-dimensional case. Further, just as the rules for evaluating derivatives in a finite-dimensional space considerably simplify the formulation of conditions for an extremum, the rules which we shall present, for constructing the sets $M(x_o)$, for functionals obtained as the result of some operations on other convex functionals, permit us to formulate conditions for an extremum for these complicated functionals.

Before beginning to consider operations which involve

$M(x_o)$, we prove the following Lemma.

Lemma 1.1. *If M_1 and M_2 are convex, weak* closed sets in B^*, and if, for every $e \in B$,*

$$\sup_{x^* \in M_1} x^*(e) = \sup_{x^* \in M_2} x^*(e),$$

then $M_1 = M_2$.

Proof. Suppose that there exists an $x_o^* \in M_1$ such that $x_o^* \notin M_2$. The set M_2 is regularly convex. Thus, there exists an e_o such that

$$\sup_{x^* \in M_2} x^*(e_o) < x_o^*(e_o) \leq \sup_{x^* \in M_1} x^*(e_o).$$

But this contradicts the assumptions of the lemma. Therefore, $M_2 \supset M_1$. Similarly, it can be shown that $M_2 \subset M_1$, which implies that $M_1 = M_2$, as was to be proved. |||

Theorem 1.3. *If $\mu(x) = c_1\mu_1(x) + c_2\mu_2(x)$ for some $c_1 \geq 0$ and $c_2 \geq 0$, and if $\mu_1(x)$ and $\mu_2(x)$ are bounded convex functionals, then $\mu(x)$ is a bounded convex functional and*

$$M(x_o) = c_1 M_1(x_o) + c_2 M_2(x_o), \tag{1.7}$$

where $M(x_o)$, $M_1(x_o)$ and $M_2(x_o)$ are the sets of support functionals to $\mu(x)$, $\mu_1(x)$, and $\mu_2(x)$, respectively.

Proof. We have

$$\frac{\partial\mu}{\partial e} = c_1 \frac{\partial\mu_1}{\partial e} + c_2 \frac{\partial\mu_2}{\partial e} = c_1 \max_{x_1^* \in M_1(x_o)} x_1^*(e) + c_2 \max_{x_2^* \in M_2(x)} x_2^*(e)$$

$$= \max_{x^* \in c_1 M_1(x_o) + c_2 M_2(x_o)} x^*(e). \qquad (1.8)$$

It is easy to verify that $\mu(x)$ is convex and bounded. Therefore,

$$\frac{\partial\mu}{\partial e} = \max_{x^* \in M(x_o)} x^*(e). \qquad (1.9)$$

The set $M(x_o)$ is convex and weak* closed by virtue of Theorem 1.1. Since the sets $M_1(x_o)$ and $M_2(x_o)$ are weak* closed and bounded, they are (Theorem 1 of the introduction) weak* compact. Therefore, the set $c_1M_1(x_o) + c_2M_2(x_o)$ is weak* closed (Theorem 2 of the introduction). The required result now follows from (1.8), (1.9), and Lemma 1.1. |||

<u>Theorem</u> 1.4. <u>Let</u> I <u>be</u> <u>a</u> <u>finite</u> <u>index</u> <u>set</u>, <u>and</u> <u>let</u> $\mu_i(x)$ <u>be</u> <u>a</u> <u>convex</u> <u>bounded</u> <u>functional</u> <u>for</u> <u>each</u> $i \in I$.

<u>Then</u> <u>the</u> <u>set</u> $M(x_o)$, <u>for</u> <u>the</u> <u>functional</u>

$$\mu(x) = \max_{i \in I} \mu_i(x),$$

(<u>which</u> <u>is</u> <u>also</u> <u>a</u> <u>convex</u> <u>and</u> <u>bounded</u> <u>functional</u>), <u>is</u> <u>given</u> <u>by</u> <u>the</u> <u>formula</u>

$$M(x_o) = \left\{x^* : x^* = \sum_{i \in I(x_o)} \lambda_i x_i^*,\ x_i^* \in M_i(x_o),\ \lambda_i \geq 0,\ \sum_{i \in I(x_o)} \lambda_i = 1\right\},$$

where

$$I(x_o) = \{i : i \in I,\ \mu(x_o) = \mu_i(x_o)\},$$

and $M_i(x_o)$ is the set of support functionals to $\mu_i(x)$ at x_o for each i.

Proof. Let us prove that $\mu(x)$ is convex.

Indeed,

$$\mu(\lambda_1 x_1 + \lambda_2 x_2) = \max_{i \in I} \mu_i(\lambda_1 x_1 + \lambda_2 x_2)$$

$$\leq \max_{i \in I} [\lambda_1 \mu_i(x_1) + \lambda_2 \mu_i(x_2)] \leq \lambda_1 \max_{i \in I} \mu_i(x_1)$$

$$+ \lambda_2 \max_{i \in I} \mu_i(x_2) = \lambda_1 \mu(x_1) + \lambda_2 \mu(x_2).$$

Moreover, if each of the functionals $\mu_i(x)$ is bounded on any bounded set, then it is easy to see that $\mu(x)$ also has this property.

First of all, let us prove that the set

$$\tilde{M} = \Big\{x^* : x^* = \sum_{i \in I(x_o)} \lambda_i x_i^*,\ x_i^* \in M_i(x_o),$$

$$\lambda_i \geq 0,\ \sum_{i \in I(x_o)} \lambda_i = 1 \text{ for } i \in I\Big\}$$

is convex and weak* closed.

Let

$$A = \bigcup_{i \in I(x_o)} M_i(x_o).$$

Then $\widetilde{M} = \text{co } A$. Furthermore, the sets $M_i(x_o)$ are convex, weak* closed, and bounded. Thus, these sets are compact in the weak* topology, and $\overline{\text{co}}\, M_i(x_o) = M_i(x_o)$. By virtue of Lemma 3, we can write

$$\overline{\text{co}}\, A = \text{co}\left(\bigcup_{i \in I(x_o)} \overline{\text{co}}\, M_i(x_o) \right) = \text{co}\left(\bigcup_{i \in I(x_o)} M_i(x_o) \right) = \text{co } A.$$

This implies that $\widetilde{M} = \text{co } A$ is closed in the weak* topology, since $\overline{\text{co}}\, A$ is weak* closed.

For each $\lambda \geq 0$, let i_λ denote an arbitrary index in the set $I(x_o + \lambda e)$. Since I is a finite set, there exists a sequence $\{\lambda_j\}$ of positive numbers converging to zero such that $i_{\lambda_j} = i^o$ for all j. We shall show that $i^o \in I(x_o)$.

Indeed, if $i^o \notin I(x_o)$, then $\mu_{i^o}(x_o) < \mu(x_o)$ and, because $\mu(x_o + \lambda e)$ and $\mu_i(x_o + \lambda e)$ are continuous with respect to λ,

$$\mu_{i^o}(x_o + \lambda_i e) < \mu(x_o + \lambda_j e)$$

for λ_j sufficiently small. This contradicts the fact that $i^o \in I(x_o + \lambda_j e)$.

Next, the equation

$$\frac{\mu(x_o + \lambda_j e) - \mu(x_o)}{\lambda_j} = \frac{\mu_{i^o}(x_o + \lambda_j e) - \mu_{i^o}(x_o)}{\lambda_j}$$

implies that $\dfrac{\partial \mu}{\partial e} = \dfrac{\partial \mu_{i^o}}{\partial e}$. For any $i \in I(x_o)$,

$$\frac{\mu(x_o + \lambda e) - \mu(x_o)}{\lambda} \geq \frac{\mu_i(x_o + \lambda e) - \mu_i(x_o)}{\lambda}$$

by definition of $\mu(x)$ and $I(x_o)$. Thus, $\frac{\partial \mu}{\partial e} \geq \frac{\partial \mu_i}{\partial e}$ for all $i \in I(x_o)$. From this, we can conclude that

$$\frac{\partial \mu}{\partial e} = \max_{x^* \in M(x_o)} x^*(e) = \max_{i \in I(x_o)} \frac{\partial \mu_i}{\partial e} .$$

But

$$\max_{i \in I(x_o)} \frac{\partial \mu_i}{\partial e} = \max_{\sum_{i \in I(x_o)} \lambda_i = 1,\ \lambda_i \geq 0} \sum_{i \in I(x_o)} \lambda_i \frac{\partial \mu_i}{\partial e}$$

$$= \max_{\sum_{i \in I(x_o)} \lambda_i = 1,\ \lambda_i \geq 0} \sum_{i \in I(x_o)} \lambda_i \max_{x_i^* \in M_i(x_o)} x_i^*(e)$$

$$= \max_{\sum_{i \in I(x_o)} \lambda_i = 1} \max_{x_i^* \in M_i(x_o)} \sum_{i \in I(x_o)} \lambda_i x_i^*(e) = \max_{x^* \in \widetilde{M}} x^*(e).$$

Therefore,

$$\max_{x^* \in M(x_o)} x^*(e) = \max_{x^* \in \widetilde{M}} x^*(e).$$

An application of Lemma 1.1 now finishes the proof of the theorem. |||

Now let B and B_1 be two Banach spaces, and let A be a bounded linear operator from B into B_1. Let $\mu_o(y)$ be a convex bounded functional defined on B_1. Then it is easy to see that the functional

$$\mu(x) = \mu_0(Ax)$$

is defined and convex on B.

Theorem 1.5. *If*

$$\mu(x) = \mu_0(Ax),$$

then the set of support functionals to $\mu(x)$ *is given by the formula*

$$M(x_0) = A^* M_0(y_0),$$

where $y_0 = Ax_0$.

Proof. First of all, let us note that the functional $\mu(x)$ is bounded on any bounded set, since $\mu_0(y)$ and A are bounded. Thus, formula (1.3) holds for $\mu(x)$.

Second of all,

$$\frac{\mu(x_0 + \lambda e) - \mu(x_0)}{\lambda} = \frac{\mu_0(Ax_0 + \lambda Ae) - \mu_0(Ax_0)}{\lambda} \xrightarrow[\lambda \to +0]{} \frac{\partial \mu_0(Ax_0)}{\partial(Ae)}$$

$$= \max_{y^* \in M_0(y_0)} y^*(Ae).$$

Therefore,

$$\frac{\partial \mu}{\partial e} = \max_{y^* \in M_0(y_0)} A^* y^*(e) = \max_{x^* \in A^* M_0(y_0)} x^*(e).$$

By virtue of Lemma 1.1, this yields the desired result. |||

Theorem 1.6. *Let* M *be a bounded, convex, weak**

closed set in B^*, and let

$$\mu(x) = \max_{x^* \in M} x^*(x).$$

Then

$$M(x_o) = \{x^* : x^* \in M,\ x^*(x_o) = \mu(x_o)\}.$$

Proof. Let $x_o^* \in M$ be such that $x_o^*(x_o) = \mu(x_o)$. Then, by definition of $\mu(x)$,

$$\mu(x) - \mu(x_o) \geq x_o^*(x) - x_o^*(x_o) = x_o^*(x - x_o),$$

i.e., $x_o^* \in M(x_o)$.

Conversely, let $x_o^* \in M(x_o)$, i.e., suppose that

$$\mu(x) - \mu(x_o) \geq x_o^*(x - x_o). \tag{1.10}$$

We shall prove that $x_o^* \in M$. Suppose the contrary. Since M is convex and weak* closed, it is regularly convex. Therefore, there exists an element e such that

$$\sup_{x^* \in M} x^*(e) < x_o^*(e). \tag{1.11}$$

On the other hand, since the maximum of a difference is not less than the difference of the maxima,

$$\max_{x^* \in M} x^*(x - x_o) \geq \mu(x) - \mu(x_o) \geq x_o^*(x - x_o).$$

Setting here $x - x_o = e$, we obtain

$$\max_{x^* \in M} x^*(e) \geq x_o^*(e).$$

But this contradicts (1.11). Thus, $x_o^* \in M$.

Let us prove that $x_o^*(x_o) = \mu(x_o)$. Assume the contrary. Since $x_o^* \in M$, this means that

$$x_o^*(x_o) < \mu(x_o).$$

Inequality (1.10) then implies that

$$\mu(x) - x_o^*(x) \geq \mu(x_o) - x_o^*(x) = \delta > 0$$

for any x. Setting $x = 0$, we obtain a contradiction. |||

CHAPTER II

CONVEX PROGRAMMING IN BANACH SPACES

In this chapter, we shall develop a theory of necessary and sufficient conditions for convex programming problems, using the properties of convex functionals which we studied in the preceding chapter.

First, let us consider the following problem. Let there be given a convex set Ω in a Banach space B and a convex bounded functional $\mu(x)$ on B. Let us determine under what conditions a point $x_o \in \Omega$ yields a minimum for $\mu(x)$ on Ω.

Let

$$\Gamma_{x_o} = \{e : e \in B, \ x_o + \lambda e \in \Omega \text{ for some } \lambda > 0\}. \qquad (2.1)$$

It is easy to verify that Γ_{x_o} is a convex cone.

Indeed, if $e \in \Gamma_{x_o}$, then there exists a $\lambda > 0$ such that $x_o + \lambda e \in \Omega$. If $e_1 = \alpha e$ with $\alpha > 0$, then $x_o + \lambda_1 e_1 \in \Omega$ for $\lambda_1 = \lambda/\alpha$. Therefore, $e_1 \in \Gamma_{x_o}$, which implies that Γ_{x_o} is a cone. Let us show that Γ_{x_o} is convex. To do this, we must show that, if

$e_1 \in \Gamma_{x_o}$ and $e_2 \in \Gamma_{x_o}$, then $e_1 + e_2 \in \Gamma_{x_o}$. Because $e_1 \in \Gamma_{x_o}$ and $e_2 \in \Gamma_{x_o}$, there exist positive numbers λ_1 and λ_2 such that

$$x_o + \lambda_1 e_1 \in \Omega \text{ and } x_o + \lambda_2 e_2 \in \Omega .$$

Since Ω is convex, for any $\alpha, \beta \geq 0$ with $\alpha + \beta = 1$, we have that

$$\alpha(x_o + \lambda_1 e_1) + \beta(x_o + \lambda_2 e_2) \in \Omega .$$

Now set

$$\alpha = \frac{\lambda_2}{\lambda_1 + \lambda_2}, \quad \beta = \frac{\lambda_1}{\lambda_1 + \lambda_2} .$$

Then

$$\alpha(x_o + \lambda_1 e_1) + \beta(x_o + \lambda_2 e_2) = x_o + \frac{\lambda_1 \lambda_2}{\lambda_1 + \lambda_2}(e_1 + e_2) \in \Omega ,$$

which precisely implies that $e_1 + e_2 \in \Gamma_{x_o}$.

The convex cone Γ_{x_o} defines a set of directions emanating from x_o with the following property. If x_o is displaced along one of these directions by a sufficiently small amount, then the displaced point remains in Ω. Indeed, if $x_o + \lambda e \in \Omega$, then the convexity of Ω implies that

$$\alpha x_o + \beta(x_o + \lambda e) \in \Omega$$

for all $\alpha, \beta \geq 0$ with $\alpha + \beta = 1$. Therefore, $x_o + \beta\lambda e \in \Omega$ for all β such that $0 \leq \beta \leq 1$, i.e., $x_o + \delta e \in \Omega$

for all δ such that $0 \leq \delta \leq \lambda$.

We shall now prove a theorem which gives necessary and sufficient conditions for a minimum of $\mu(x)$ on Ω.

Theorem 2.1. *A convex bounded functional $\mu(x)$ achieves its minimum at a point $x_o \in \Omega$, where Ω is a convex set, if and only if*

$$\Gamma^*_{x_o} \cap M(x_o) \neq \emptyset. \tag{2.2}$$

Remark. Here, $M(x_o)$ is the set of support functionals to $\mu(x)$ at x_o, and $\Gamma^*_{x_o}$ is the cone dual to Γ_{x_o}, i.e.,

$$\Gamma^*_{x_o} = \{x^* : x^* \in B^*, \ x^*(e) \geq 0 \text{ for all } e \in \Gamma_{x_o}\}.$$

It is known (Lemma 5 of the introduction) that $\Gamma^*_{x_o}$ is a convex, weak* closed cone in B^*.

Proof. Necessity. Suppose that

$$\mu(x_o) \leq \mu(x) \text{ for all } x \in \Omega,$$

and that condition (2.2) is violated. The set $\Gamma^*_{x_o} - M(x_o)$ is convex and weak* closed. Indeed, $\Gamma^*_{x_o}$ is weak* closed, and, by virtue of Theorem 1.1, $M(x_o)$ is weak* closed and bounded in the metric topology. Therefore (Theorem 1 of the introduction), $M(x_o)$ is weak* compact, and, according to Theorem 2 of the introduction, the set $\Gamma^*_{x_o} - M(x_o)$ is weak* closed.

Since $\Gamma^*_{x_o} - M(x_o)$ is convex and weak* closed, it is regularly convex.

By virtue of our assumption, the zero functional does not belong to $\Gamma^*_{x_o} - M(x_o)$. Therefore, there exists an $e_o \in B$ such that

$$\inf_{x^* \in \Gamma^*_{x_o} - M(x_o)} x^*(e_o) \geq \delta > 0,$$

or

$$\inf_{x^* \in \Gamma^*_{x_o}} x^*(e_o) \geq \delta + \max_{x^* \in M(x_o)} x^*(e_o). \qquad (2.3)$$

But, since $\Gamma^*_{x_o}$ is a cone, the left-hand side of the last inequality vanishes. Indeed, if

$$\inf_{x^* \in \Gamma^*_{x_o}} x^*(e_o) < 0,$$

then there exists a functional $x_1^* \in \Gamma^*_{x_o}$ such that

$$x_1^*(e_o) < 0.$$

Since $\Gamma^*_{x_o}$ is a cone, $\alpha x_1^* \in \Gamma^*_{x_o}$ for all $\alpha > 0$. Therefore,

$$\inf_{x^* \in \Gamma^*_{x_o}} x^*(e_o) \leq \alpha x_1^*(e_o)$$

for all $\alpha > 0$, which implies that $\inf_{x^* \in \Gamma^*_{x_o}} x^*(e_o) = -\infty$. But this contradicts Inequality (2.3) whose right-hand side is bounded. Thus,

$$\inf_{x^* \in \Gamma^*_{x_o}} x^*(e_o) = 0. \qquad (2.4)$$

Hence, $e_o \in \overline{\Gamma}_{x_o}$ (Lemma 4 of the introduction), where the bar indicates closure in the strong topology of B. Moreover, formulas (1.3), (2.3) and (2.4) imply that

$$\frac{\partial\mu(x_o)}{\partial e_o} \leq -\delta.$$

Since $e_o \in \overline{\Gamma}_{x_o}$, there exists a direction e in Γ_{x_o} such that

$$\|e - e_o\| \leq \frac{\delta}{2} \left(\sup_{x^* \in M(x_o)} \|x^*\| \right)^{-1}.$$

Therefore,

$$\frac{\partial\mu(x_o)}{\partial e} = \max_{x^* \in M(x_o)} x^*(e) \leq \max_{x^* \in M(x_o)} x^*(e_o)$$

$$+ \max_{x^* \in M(x_o)} x^*(e - e_o) \leq -\delta + \|e - e_o\| \sup_{x^* \in M(x_o)} \|x^*\| \leq -\frac{\delta}{2}.$$

Thus, for λ sufficiently small, $x_o + \lambda e \in \Omega$ and $\mu(x_o + \lambda e) \leq \mu(x_o) - \lambda \frac{\delta}{4}$, which contradicts the fact that x_o is a minimum point of $\mu(x)$ on Ω.

Sufficiency. Let Condition (2.2) hold. Then there exists a functional x^*_o such that $x^*_o \in M(x_o) \cap \Gamma^*_{x_o}$, i.e.,

$$\mu(x) - \mu(x_o) \geq x^*_o(x - x_o) \qquad (2.5)$$

for all x, and $x_o^*(e) \geq 0$ for all $e \in \Gamma_{x_o}$. But if $x \in \Omega$, then it is easy to verify that the direction $e = x - x_o$ belongs to the cone Γ_{x_o}, and thus,

$$x_o^*(x - x_o) \geq 0 \text{ for all } x \in \Omega.$$

Comparing this inequality with (2. 5), we at once see that the sufficiency also holds. |||

Corollary. A convex bounded functional $\mu(x)$ on a set Ω achieves its minimum at a point x_o if and only if there exists a functional $x_o^* \in M(x_o)$ such that

$$x_o^*(x) \geq x_o^*(x_o) \tag{2. 6}$$

for all $x \in \Omega$.

Proof. Indeed, if (2. 6) holds, then

$$\mu(x) - \mu(x_o) \geq x_o^*(x - x_o) \geq 0$$

for all $x \in \Omega$.

Conversely, if x_o is a minimum point, then, by virtue of the preceding theorem, there exists a functional $x_o^* \in M(x_o) \cap \Gamma_{x_o}^*$. This means that

$$x_o^*(x - x_o) \geq 0$$

for all $x \in \Omega$, which implies that (2. 6) holds. |||

If one wishes to consider Theorem 2. 1 to be a tool for effectively constructing necessary and sufficient conditions for a minimum, one can easily see that the functional being minimized and the set Ω have not been equally made use of in the formulation of this theorem.

We have already developed an apparatus for constructing the set $M(x_0)$, but, up to now, the cone $\Gamma^*_{x_0}$ has only been specified by its definition.

The following theorem, to a large extent, eliminates this shortcoming, and makes it possible to employ the apparatus which has been developed for the construction of $M(x_0)$ to investigate $\Gamma^*_{x_0}$.

<u>Theorem</u> 2. 2. <u>Let the set Ω be given by the inequality</u>

$$\Omega = \{x : \mu(x) \le 0\},$$

<u>where</u> $\mu(x)$ <u>is a convex, bounded functional such that, for some</u> x_1, $\mu(x_1) < 0$. <u>Then, if</u> $\mu(x_0) < 0$, $\Gamma^*_{x_0} = \{0\}$. <u>If</u> $\mu(x_0) = 0$, <u>then</u>

$$\Gamma^*_{x_0} = \{x^* : x^* = \gamma x^*_0,\ \gamma \le 0,\ x^*_0 \in M(x_0)\}.$$

<u>Proof.</u> Let us first consider the case where $\mu(x_0) < 0$. Then, for any e and for sufficiently small λ, by virtue of formula (1. 3),

$$\mu(x_0 + \lambda e) = \mu(x_0) + \lambda \max_{x^* \in M(x_0)} x^*(e) + o(\lambda) < 0.$$

Therefore, $\Gamma_{x_0} = B$, and $\Gamma^*_{x_0} = \{0\}$.

Now let $\mu(x_0) = 0$. In this case,

$$\Gamma_{x_0} = \{e : \mu(x_0 + \lambda e) \leq 0 = \mu(x_0) \text{ for some } \lambda > 0\}.$$

This implies that, if $e \in \Gamma_{x_0}$, then

$$0 \geq \mu(x_0 + \lambda e) - \mu(x_0) \geq \lambda x^*(e)$$

for any $x^* \in M(x_0)$. Thus, $\gamma x^*(e) \geq 0$ for all $\gamma \leq 0$ and all $e \in \Gamma_{x_0}$. But this means that $\gamma x^* \in \Gamma^*_{x_0}$.

Let us now prove the inclusion in the opposite direction; i.e., we shall prove that, for any $x^* \in \Gamma^*_{x_0}$, there exist a $\gamma \leq 0$ and an $x^*_0 \in M(x_0)$ such that $x^* = \gamma x^*_0$.

Assume the contrary. Then there exists a functional $\tilde{x}^* \in \Gamma^*_{x_0}$ such that $\alpha \tilde{x}^* \notin M(x_0)$ for any $\alpha < 0$. Moreover, $\alpha \tilde{x}^* \notin M(x_0)$ also for $\alpha = 0$, because

$$0 > \mu(x_1) - \mu(x_0) \geq x^*(x_1 - x_0)$$

when $x^* \in M(x_0)$. This implies that $M(x_0)$ does not contain the zero functional.

Let us consider the set $\alpha \tilde{x}^* - M(x_0)$, for any $\alpha \leq 0$. This set is convex, weak* closed (since $M(x_0)$ is weak* closed and compact) and, according to what has just been said, does not contain the zero functional. Therefore, there exists an $e \in B$ such that

$$\alpha\tilde{x}^*(e) \geq 0 > x^*(e)$$

for all $\alpha \leq 0$ and $x^* \in M(x_o)$. But then

$$\frac{\partial\mu(x_o)}{\partial e} = \max_{x^* \in M(x_o)} x^*(e) < 0,$$

and thus, $e \in \Gamma_{x_o}$. Moreover, it is easy to see that e is an interior point of Γ_{x_o}, and, therefore [4], $x^*(e) > 0$ for all $x^* \in \Gamma^*_{x_o}$, which means that $\tilde{x}^*(e) > 0$. But this contradicts the fact that $\alpha\tilde{x}^*(e) \geq 0$ for all $\alpha \leq 0$.

The contradiction which we have just obtained shows that every functional $x^* \in \Gamma^*_{x_o}$ can be represented in the form $x^* = \gamma x^*_o$ with $\gamma \leq 0$ and $x^*_o \in M(x_o)$. Previously, it was shown that $\gamma x^*_o \in \Gamma^*_{x_o}$ for any $\gamma \leq 0$ and $x^*_o \in M(x_o)$. This implies that the theorem is true. |||

We shall now derive necessary and sufficient conditions for optimality in the case where the set Ω is given by various collections of equalities and inequalities.

Theorem 2.3. A convex bounded functional $\mu(x)$ attains its minimum on the entire space B at a point x_o if and only if

$$0 \in M(x_o).$$

Proof. Since the set Ω in this case coincides with the entire space, $\Gamma_{x_o} = B$, $\Gamma^*_{x_o} = \{0\}$, and the result follows immediately from Theorem 2.1. |||

Theorem 2.4. *Suppose that a set* Ω *is specified by the collection of inequalities*

$$\mu_i(x) \leq 0 \text{ for } i = 1, \ldots, n,$$

where, for each i, $\mu_i(x)$ *is a convex bounded functional, and also suppose that there exists a point* x_1 *such that* $\mu_i(x_1) < 0$ *for each* $i = 1, \ldots, n$. *Then the convex bounded functional* $\mu_o(x)$ *achieves its minimum on* Ω *at a point* x_o *if and only if there exist elements* $x_i^* \in M_i(x_o)$ *and numbers* λ_i, $i = 0, 1, \ldots, n$, *such that*

$$x_o^* = \sum_{i=1}^{n} \lambda_i x_i^*, \quad \lambda_i \mu_i(x_o) = 0 \text{ and } \lambda_i \leq 0 \text{ for } i = 1, \ldots, n.$$

Proof. Let us introduce the functional

$$\mu(x) = \max_{1 \leq i \leq n} \mu_i(x).$$

Then Ω can be specified with the aid of a single inequality; namely,

$$\Omega = \{x : \mu(x) \leq 0\}.$$

If $\mu(x_o) < 0$, then our desired result follows from Theorems 2.1 and 2.2, because, in this case, $\Gamma^*_{x_o} = \{0\}$ and $M(x_o)$ must contain the zero functional, and the conditions of the theorem are satisfied with $\lambda_i = 0$ for $i = 1, \ldots, n$.

Suppose that $\mu(x_o) = 0$. Then, by Theorem 2.2, $\Gamma^*_{x_o} = \{\gamma\bar{x}^* : \gamma \leq 0, \ \bar{x}^* \in M(x_o)\}$. On the other hand, on the basis of Theorem 1.4, $\bar{x}^* \in M(x_o)$ if and only if $\bar{x}^*$ can be represented in the form

$$\bar{x}^* = \sum_{i \in I(x_o)} \bar{\lambda}_i x_i^*,$$

where $\bar{\lambda}_i \geq 0$ for each i, $\sum_{i \in I(x_o)} \bar{\lambda}_i = 1$, $x_i^* \in M_i(x)$ for each i, and $I(x_o) = \{i : \mu_i(x_o) = \mu(x_o) = 0\}$.

Thus, $x^* \in \Gamma^*_{x_o}$ if and only if it can be written in the form

$$x^* = \gamma \sum_{i \in I(x_o)} \bar{\lambda}_i x_i^* \text{ with } \gamma \leq 0.$$

On the basis of Theorem 2.1, a minimum is achieved at a point $x_o \in \Omega$ only if there exists a functional $x_o^* \in M_o(x_o) \cap \Gamma^*_{x_o}$. But, from what has just been said, it follows that this functional can be written in the form

$$x_o^* = \gamma \sum_{i \in I(x_o)} \bar{\lambda}_i x_i^*.$$

Let $\lambda_i = \gamma\bar{\lambda}_i$ for $i \in I(x_o)$, and let $\lambda_i = 0$ for $i \notin I(x_o)$. Then

$$x_o^* = \sum_{i=1}^{n} \lambda_i x_i^*,$$

where, for each i, $\lambda_i \leq 0$ (since $\gamma \leq 0$ and $\bar{\lambda}_i \geq 0$). In addition, $\lambda_i\mu_i(x_o) = 0$ for each i, because, if $i \in I(x_o)$, then $\mu_i(x_o) = 0$, and if $i \notin I(x_o)$, then $\lambda_i = 0$. |||

Theorem 2. 4 yields necessary and sufficient conditions for a minimum in a convex programming problem when the set Ω is given by a collection of inequalities. In order to obtain the conditions for a minimum in the case where Ω is given by a collection of equalities and inequalities, it is most convenient to employ a Kuhn-Tucker theorem. A proof of this theorem can be found in [11]. We state this theorem here for the sake of completeness. Later, we shall show how to obtain this theorem as a corollary of the more general Theorem 4. 1 of Chapter 4.

<u>Theorem</u> 2. 5 <u>(Kuhn-Tucker)</u>. <u>Let</u> $\mu_i(x)$, $i = -m, \ldots, -1, 0, 1, \ldots, n$, <u>be functionals defined on a linear space</u> E <u>such that the</u> $\mu_i(x)$ <u>are convex for</u> $i \leq 0$ <u>and linear for</u> $i > 0$. <u>Then, in order that</u> $\mu_o(x)$ <u>achieves its minimum at a point</u> $x_o \in E$, <u>subject to the constraints</u>

$$\mu_i(x) \leq 0 \text{ for } i = -m, \ldots, -1,$$

$$\mu_i(x) = 0 \text{ for } i = 1, \ldots, n,$$

$$x \in X,$$

<u>where</u> X <u>is some given convex set in</u> E, <u>it is necessary that</u>

there exist constants λ_i, $i = -m, \ldots, n$, *such that*

$$\sum_{i=-m}^{n} \lambda_i \mu_i(x_o) \leq \sum_{i=-m}^{n} \lambda_i \mu_i(x)$$

for all $x \in X$. *Moreover,* $\lambda_i \geq 0$ *for each* $i \leq 0$, *and* $\lambda_i \mu_i(x_o) = 0$ *for each* $i \neq 0$. *If* $\lambda_o > 0$, *then the conditions are also sufficient.*

This Kuhn-Tucker theorem has a global character. The following theorem gives necessary and sufficient conditions in differential form. Such a form is often particularly convenient from a practical standpoint.

Theorem 2.6. *Let* B *be a Banach space. In order that a point* x_o *be a solution to the minimization problem*

$$\min \mu_o(x),$$

$$\mu_i(x) \leq 0 \text{ for } -m \leq i \leq -1,$$

$$\mu_i(x) = 0 \text{ for } 1 \leq i \leq n,$$

$$x \in X,$$

where the $\mu_i(x)$ *are convex for* $i \leq 0$ *and linear for* $i > 0$ *and* X *is a convex subset of* B, *it is necessary that there exist constants* λ_i *and functionals* $x_i^* \in M_i(x_o)$ $(i = -m, \ldots, n)$ *such that*

$$\sum_{i=-m}^{n} \lambda_i x_i^*(x_o) \leq \sum_{i=-m}^{n} \lambda_i x_i^*(x) \text{ for all } x \in X,$$

<u>and</u>

$$\lambda_i \geq 0 \text{ and } \lambda_i \mu_i(x_o) = 0 \text{ for each } i < 0, \ \lambda_o \geq 0.$$

<u>If</u> $\lambda_o > 0$, <u>then these conditions are also sufficient.</u>

<u>Proof.</u> The theorem is an immediate consequence of Theorem 2.5, the corollary of Theorem 2.1, and Theorem 1.3. |||

CHAPTER III

QUASI-DIFFERENTIABLE FUNCTIONALS

It is easy to see that, in order to prove almost all of the results of Chapter I which were concerned with methods of evaluating directional differentials, the convexity of the functional under consideration was not explicitly used. All of the reasoning was based on the formula (see (1. 3))

$$\frac{\partial \mu(x_o)}{\partial e} = \max_{x^* \in M(x_o)} x^*(e).$$

It is useful to introduce the following definition.

Definition 3. 1. A functional is said to be quasi-differentiable at a point x_o if there exists a convex, weak* closed set $M(x_o)$ such that formula (1. 3) holds.

Let us discuss how broad the class of quasi-differentiable functionals is. First of all, it is clear that this class contains all Gâteaux-differentiable functionals, since, in this case, we can take for $M(x_o)$ the set consisting of the single functional x^*_o, the Gâteaux differential of $\mu(x)$ at x_o. Second of all, the results of Chapter I show that all convex functionals are quasi-differentiable.

Further, Theorems 1.3 and 1.4 still hold if, in their statements, the words "convex functional" are replaced by "quasi-differentiable functional". Therefore, the operations of forming linear combinations with positive coefficients and of taking the maximum of a finite number of functionals do not lead us out of the class of quasi-differentiable functionals.

The following theorem is a generalization of Theorem 1.5.

Theorem 3.1. *Let* B_o *and* B_1 *be Banach spaces, let* $\mu(y)$ *be a quasi-differentiable functional which also satisfies a Lipschitz condition, and let* $A(x) : B_o \to B_1$ *be a Gâteaux-differentiable operator. Then*

$$\mu_o(x) = \mu(A(x))$$

is a quasi-differentiable functional on B_o, *and*

$$M_o(x_o) = (A'_o)^* M(A(x_o)),$$

where A'_o *is the Gâteaux differential of* $A(x)$ *at* x_o.

Proof.

$$\frac{\mu_o(x_o + \lambda e) - \mu_o(x_o)}{\lambda} = \frac{\mu(A(x_o + \lambda e)) - \mu(A(x_o))}{\lambda}$$

$$= \frac{\left[\mu(A(x_o) + \lambda A_o' e + r(\lambda)) - \mu(A(x_o) + \lambda A_o' e)\right]}{\lambda}$$

$$+ \frac{\mu(A(x_o) + \lambda A_o' e) - \mu(A(x_o))}{\lambda},$$

$$\text{where } \lim_{\lambda \to 0} \frac{r(\lambda)}{\lambda} = 0.$$

As $\lambda \to +0$, the right-hand side of this equation tends to

$$\frac{\partial \mu(A(x_o))}{\partial(A_o' e)} = \max_{y^* \in M(A(x_o))} y^*(A_o' e).$$

This implies the desired conclusion. |||

Let us now show that, in a broad class of cases, the operation consisting of taking the maximum with respect to a parameter does not lead out of the class of quasi-differentiable functionals. We begin by proving some auxiliary assertions.

Let $\varphi(x, \alpha)$ be a functional which is continuous with respect to x and α, where x belongs to a normed space B, and α is an element in a compact topological space Z. Let us construct the functional

$$\mu(x) = \max_{\alpha \in Z} \varphi(x, \alpha),$$

and let

$$Z(x) = \{\alpha : \alpha \in Z,\ \varphi(x,\alpha) = \mu(x)\}.$$

Lemma 3.1. The functional $\mu(x)$ is continuous.

Proof. We have

$$\varphi(x,\alpha) = \mu(x) \geq \varphi(x,\alpha_o),$$

$$\varphi(x_o,\alpha) \leq \mu(x_o) = \varphi(x_o,\alpha_o)$$

$$\alpha \in Z(x),\quad \alpha_o \in Z(x_o).$$

Subtracting these inequalities, we obtain

$$\varphi(x,\alpha) - \varphi(x_o,\alpha) \geq \mu(x) - \mu(x_o) \geq \varphi(x,\alpha_o) - \varphi(x_o,\alpha_o). \qquad (3.1)$$

The right-hand side of this inequality tends to zero as $x \to x_o$. By virtue of the compactness of Z and the continuity of $\varphi(x,\alpha)$ (with respect to its arguments), the left-hand side of (3.1) also tends to zero.

Indeed, let us suppose that the difference

$$\varphi(x,\alpha) - \varphi(x_o,\alpha) \text{ where } \alpha \in Z(x),$$

does not tend to zero as $x \to x_o$. This means that there exists a sequence $\{(x_i,\alpha_i)\}$, where $x_i \to x_o$ and $\alpha_i \in Z(x_i)$ for each i, such that

$$|\varphi(x_i,\alpha_i) - \varphi(x_o,\alpha_i)| \geq \delta > 0.$$

Since Z is compact, we can choose a convergent subsequence

of the sequence $\{\alpha_i\}$. Without loss of generality, we shall assume that the original sequence $\{\alpha_i\}$ itself converges to $\alpha_o \in Z$.

Passing to the limit in the last inequality, we obtain, by virtue of the continuity of $\varphi(x, \alpha)$, that

$$0 \geq \delta > 0,$$

which is absurd and thus shows that the left-hand side of (3.1) tends to zero, completing the proof of the lemma. |||

<u>Lemma</u> 3.2. <u>For every neighborhood</u> $\omega \subset Z$ <u>of the set</u> $Z(x_o)$, <u>there exists a number</u> δ <u>such that</u>

$$Z(x) \subset \omega$$

<u>whenever</u> $\|x - x_o\| \leq \delta$.

We shall argue by contradiction. Thus, suppose that, for some neighborhood ω_o of $Z(x_o)$, there exist a sequence $x_i \to x_o$, and, for every $i = 1, 2, \ldots, a$, point $\alpha_i \in Z(x_i)$ such that $\alpha_i \notin \omega_o$. Since Z is compact, we can assume that $\alpha_i \to \alpha_o$.

By definition,

$$\mu(x_i) = \varphi(x_i, \alpha_i).$$

Passing to the limit, we obtain that

$$\mu(x_o) = \varphi(x_o, \alpha_o),$$

i.e., that $\alpha_o \in Z(x_o)$. Now $\alpha_i \to \alpha_o$, $\alpha_i \notin \omega_o$ for each i, and

$\alpha_o \in Z(x_o) \subset \omega_o$. But, since ω_o is open, this means that $\alpha_i \in \omega_o$ for sufficiently large i, which is a contradiction. |||

Now let us investigate the differential properties of the functional $\mu(x)$. Let $x(\lambda) = x_o + \lambda e$. Substituting this into (3.1), and dividing this inequality by $\lambda > 0$, we obtain

$$\frac{\varphi(x(\lambda), \alpha) - \varphi(x_o, \alpha)}{\lambda} \geq \frac{\mu(x(\lambda)) - \mu(x_o)}{\lambda}$$

$$\geq \frac{\varphi(x(\lambda), \alpha_o) - \varphi(x_o, \alpha_o)}{\lambda}, \text{ for all } \alpha \in Z(x(\lambda)). \qquad (3.2)$$

From the left-hand side of this inequality we obtain that, if $\varphi(x, \alpha)$ is differentiable in the direction e, then

$$\underline{\lim}_{\lambda \to +0} \frac{\mu(x(\lambda)) - \mu(x_o)}{\lambda} \geq \frac{\partial \varphi(x_o, \alpha_o)}{\partial e},$$

or, since α_o is an arbitrary element of $Z(x_o)$,

$$\underline{\lim}_{\lambda \to +0} \frac{\mu(x(\lambda)) - \mu(x_o)}{\lambda} \geq \sup_{\alpha \in Z(x_o)} \frac{\partial \varphi(x_o, \alpha)}{\partial e}. \qquad (3.3)$$

Now suppose that $\varphi(x, \alpha)$ is differentiable in the direction e at the point x_o, i.e., that, for $\lambda > 0$,

$$\varphi(x_o + \lambda e, \alpha) = \varphi(x_o, \alpha) + \lambda \frac{\partial \varphi(x_o, \alpha)}{\partial e} + \lambda \gamma(\lambda, \alpha)$$

and, in addition, that $\gamma(\lambda, \alpha) \to 0$ as $\lambda \to +0$ uniformly with respect to α. Then the directional derivative $\frac{\partial\varphi(x_o, \alpha)}{\partial e}$, as the uniform limit of the continuous functions

$$\frac{\varphi(x_o + \lambda e, \alpha) - \varphi(x_o, \alpha)}{\lambda} \quad \text{and} \quad \inf_{\omega \supset Z(x_o)} \sup_{\alpha \in \omega} \frac{\partial\varphi(x_o, \alpha)}{\partial e}$$

$$= \sup_{\alpha \in Z(x_o)} \frac{\partial\varphi(x_o, \alpha)}{\partial e},$$

is continuous in α.

By virtue of Lemma 3.2, we obtain from (3.2) that

$$\frac{\mu(x(\lambda)) - \mu(x_o)}{\lambda} \le \sup_{\alpha \in \omega} \left[\frac{\partial\varphi(x_o, \alpha)}{\partial e} + \gamma(\lambda, \alpha) \right]$$

for any $\omega \supset Z(x_o)$ and for sufficiently small λ. Passing to the limit as $\lambda \to +0$, we conclude that

$$\overline{\lim_{\lambda \to +0}} \frac{\mu(x(\lambda)) - \mu(x_o)}{\lambda} \le \sup_{\alpha \in \omega} \frac{\partial\varphi(x_o, \alpha)}{\partial e}.$$

Because ω is an arbitrary neighborhood of $Z(x_o)$, we finally obtain

$$\overline{\lim_{\lambda \to +0}} \frac{\mu(x(\lambda)) - \mu(x_o)}{\lambda} \le \inf_{\omega \supset Z(x_o)} \sup_{\alpha \in \omega} \frac{\partial\varphi(x_o, \alpha)}{\partial e}$$

$$= \sup_{\alpha \in Z(x_o)} \frac{\partial\varphi(x_o, \alpha)}{\partial e}. \tag{3.4}$$

A comparison of (3.3) and (3.4) leads us to the following

theorem.

Theorem 3.2. *Let $\varphi(x, \alpha)$ be a functional which is continuous in x and α, where $x \in B$ and $\alpha \in Z$ (Z is a compact topological space).*

Moreover, let

$$\varphi(x_o + \lambda e, \alpha) = \varphi(x_o, \alpha) + \lambda \frac{\partial \varphi(x_o, \alpha)}{\partial e} + \lambda \gamma(x, \alpha) \text{ for } \lambda > 0,$$

where $\gamma(\lambda, \alpha) \to 0$ uniformly in α as $\lambda \to +0$. Then, at the point x_o, the functional

$$\mu(x) = \max_{\alpha \in Z} \varphi(x, \alpha)$$

is differentiable in the direction e, and

$$\frac{\partial \mu(x_o)}{\partial e} = \sup_{\alpha \in Z(x_o)} \frac{\partial \varphi(x_o, \alpha)}{\partial e}. \tag{3.5}$$

Remark 1. It is easy to see that the hypothesis that φ is continuous in x and α can, without any harm, be replaced by the hypothesis that the function $\varphi(x_o + \lambda e, \alpha)$ is continuous in λ and α.

Remark 2. If $\varphi(x, \alpha)$ is a functional which is convex in x for each fixed α, then the theorem holds if $\frac{\partial \varphi(x_o, \alpha)}{\partial e}$ depends continuously on α.

Indeed, a convex functional is always directionally

differentiable, and the quotient (for fixed α)

$$\frac{\varphi(x_o + \lambda e, \alpha) - \varphi(x_o, \alpha)}{\lambda}$$

tends to $\frac{\partial\varphi(x_o, \alpha)}{\partial e}$, decreasing monotonically [6] (also see Lemma 6 in the introduction). But then, by Dini's theorem [45], $\gamma(\lambda, \alpha)$ tends to zero uniformly with respect to α.

Theorem 3.3. *If $\varphi(x_o + \lambda e, \alpha)$ is a functional which is continuous in λ and α, and, moreover, if the differential $\frac{\partial\varphi(x_o + \lambda e, \alpha)}{\partial\lambda}$ exists and is continuous in $\lambda \in [0, 1]$ and $\alpha \in Z$, then*

$$\frac{\partial\mu(x_o)}{\partial e} = \max_{\alpha \in Z(x_o)} \frac{\partial\varphi(x, \alpha)}{\partial e}. \tag{3.6}$$

This may be proved analogously to the way in which we proved the preceding theorem, with one exception: in order to establish Inequality (3.4), it is necessary to employ the mean-value theorem of the differential calculus to obtain the formula

$$\frac{\varphi(x_o + \lambda e, \alpha) - \varphi(x_o, \alpha)}{\lambda} = \frac{\partial\varphi(x_o + \xi(\lambda)e, \alpha)}{\partial\lambda},$$

where $0 \leq \xi(\lambda) \leq \lambda$, and to note that

$$\left.\frac{\partial\varphi(x_o + \lambda e, \alpha)}{\partial\lambda}\right|_{\lambda = 0} = \frac{\partial\varphi(x_o, \alpha)}{\partial e}.$$

Theorem 3.4. *If, for every* α, $\varphi(x,\alpha)$ *is a functional which is quasi-differentiable at* x_o, *and if the hypotheses of Theorem 3.2 are satisfied for every* e, *then also* $\mu(x)$ *is a functional which is quasi-differentiable at* x_o. *Moreover,*

$$M(x_o) = \overline{\text{co}} \bigcup_{\alpha \in Z(x_o)} M(x_o, \alpha), \tag{3.7}$$

where $M(x_o, \alpha)$ *is the set of support functionals to* $\varphi(x,\alpha)$ *at* x_o, $M(x_o)$ *is the set of support functionals to* $\mu(x)$ *at* x_o, *and* $\overline{\text{co}}$ A *denotes the weak* closure of the convex hull of* A.

Proof. On the basis of Theorem 3.2 and the definition of a quasi-differentiable functional, we obtain

$$\frac{\partial \mu(x_o)}{\partial e} = \sup_{\alpha \in Z(x_o)} \frac{\partial \varphi(x_o, \alpha)}{\partial e} = \sup_{\alpha \in Z(x_o)} \sup_{x^* \in M(x_o, \alpha)} x^*(e). \tag{3.8}$$

But, as is easily seen, the maximum in the right-hand side of this equation equals the maximum of $x^*(e)$ as x^* ranges over the set $M(x_o)$ defined by Formula (3.7). Indeed, $M(x_o)$ is the weak* closure of the set

$$\{x^* : x^* = \sum \lambda_i x_i^*,\ \sum \lambda_i = 1,\ x_i^* \in M(x_o, \alpha_i) \text{ and } \lambda_i \geq 0 \text{ for all } i\}.$$

Therefore,

$$x^*(e) = \sum \lambda_i x_i^*(e) \leq \max_i x_i^*(e) \leq \sup_{\alpha \in Z(x_o)} \sup_{x^* \in M(x_o, \alpha)} x^*(e),$$

and, thus, by virtue of the fact that the supremum of the function $x^*(e)$ (which is continuous in its argument x^*) on some set and on the closure of this set coincide,

$$\sup_{x^* \in M(x_0)} x^*(e) \leq \sup_{\alpha \in Z(x_0)} \sup_{x^* \in M(x_0, \alpha)} x^*(e).$$

But, on the other hand, $M(x_0) \supset M(x_0, \alpha)$ for any $\alpha \in Z(x_0)$, and thus,

$$\sup_{x^* \in M(x_0)} x^*(e) \geq \sup_{\alpha \in Z(x_0)} \sup_{x^* \in M(x_0, \alpha)} x^*(e).$$

Hence, the indicated maxima coincide, and we can write

$$\frac{\partial \mu(x_0)}{\partial e} = \sup_{x^* \in M(x_0)} x^*(e).$$

To finish the proof of the theorem, it remains to note that, by definition, $M(x_0)$ is convex and weak* closed. |||

The representation of $M(x_0)$ in the form (3.7) is not always effective. We shall now indicate a particular case in which $M(x_0)$ can be described more effectively.

Theorem 3.5. Let $\varphi(x, \alpha)$ be a functional which is Gâteaux differentiable with respect to x and continuous with respect to α. We shall suppose that α ranges over a compact metric space Z. Further, we suppose that $x^*_{x,\alpha}$ (which denotes the Gâteaux differential of φ with respect to x, *evaluated at (x, α)) is weak* continuous when considered as a function of x and α. Then*

$$\mu(x) = \max_{\alpha \in Z} \varphi(x, \alpha)$$

is a quasi-differentiable functional. Moreover, $M(x_o)$ consists of all functionals x^* which can be represented in the form

$$x^*(e) = \int_{Z(x_o)} x^*_{x_o, \alpha}(e)\, \nu(d\alpha), \tag{3.9}$$

where ν is a non-negative measure whose total variation on $Z(x_o)$ equals 1.

Proof. Since the function $\varphi(x_o + \lambda e, \alpha)$ has the following derivative with respect to λ

$$\frac{\partial \varphi}{\partial \lambda} = x^*_{x_o + \lambda e, \alpha}(e),$$

and this derivative is, by virtue of the assumptions of the theorem, continuous in λ and α, then, on the basis of Theorem 3.3, we can conclude that, for every e,

$$\frac{\partial \mu(x_o)}{\partial e} = \max_{\alpha \in Z(x_o)} \frac{\partial \varphi(x_o, \alpha)}{\partial e} = \max_{\alpha \in Z(x_o)} x^*_\alpha(e),$$

where we have denoted $x^*_{x_o, \alpha}$ by x^*_α.

By definition of an integral with respect to a measure,

and by some of its standard properties, we conclude that, for any e,

$$\max_{\alpha \in Z(x_o)} x^*_\alpha(e) = \max_\nu \int_{Z(x_o)} x^*_\alpha(e)\nu(d\alpha),$$

where the maximum is taken over all measures with the properties indicated in the theorem statement.

Thus,

$$\frac{\partial \mu(x_o)}{\partial e} = \max_\nu \int_{Z(x_o)} x^*_\alpha(e)\nu(d\alpha).$$

To finish the proof, we need only convince ourselves that the set of all functionals which can be represented in the form (3.9) is convex, weak* closed and bounded. But the convexity is obvious.

Let us show that this set is weak* closed. In order to do this, we note that each of the measures under consideration can be thought of as an element of the unit ball (the total variation of ν is 1) in the space $C^*(Z)$, i.e., in the space conjugate to the space of all functions continuous on Z. Since Z is compact, C(Z) is separable. But, in this case, the weak* topology on the unit ball in C^* can be defined by means of some metric (see [1, Theorem V.5.1, p. 426]).

Since the notions of compactness and sequential

compactness coincide for metric spaces, then, because the unit ball of $C^*(Z)$ is weak* compact (Theorem 1 of the introduction), this ball is also weak* sequentially compact. Now, suppose that a sequence of functionals

$$x_n^*(e) = \int_{Z(x_o)} x_\alpha^*(e)\nu_n(d\alpha) \qquad (3.10)$$

weak* converges to x^*. Let us choose from the sequence ν_n a subsequence which is weak* convergent (in the sense of the space of continuous functions $C(Z)$). Without loss of generality, we can assume that the sequence ν_n itself weak* converges to ν_o. Then, because for every set E of the Borel σ - algebra of Z,

$$\nu_o(E) = \int_E \nu_o(d\alpha) = \lim_{n\to\infty} \int_E \nu_n(d\alpha) = \lim_{n\to\infty} \nu_n(E),$$

we conclude that ν_o is a non-negative measure, and that $\nu_o(Z(x_o)) = 1$.

Passing to the limit in (3.10), we obtain that

$$x^*(e) = \int_{Z(x_o)} x_\alpha^*(e)\nu_o(d\alpha),$$

which proves that the set under consideration is weak* closed. |||

CHAPTER IV

NECESSARY CONDITIONS FOR AN EXTREMUM IN GENERAL MATHEMATICAL PROGRAMMING PROBLEMS

In this chapter we shall derive necessary conditions for an extremum which generalize the usual Lagrange multiplier rule to a broad class of problems.

Let L be a linear space, let φ_i, for $i = -m, \ldots, 0, \ldots, k$, denote functionals defined on this space, and let M be some set. Consider the problem:

$$\begin{aligned} &\min \varphi_o(x), \\ &\varphi_i(x) \leq 0 \text{ for } i < 0, \\ &\varphi_i(x) = 0 \text{ for } i > 0, \\ &x \in M. \end{aligned} \tag{4.1}$$

Since such a problem formulation is very general, one has to make some basic assumptions in order to obtain effective conditions for an extremum.

Let x_o be a solution to the problem. Then we shall suppose the following:

IV. Necessary Conditions for Extremum in Mathematical Programming

1. There exists a convex cone K_M such that, if $e \in K_M$, then

$$x(\lambda) = x_o + \lambda e + \sum_{i=1}^{k} r_i(\lambda) e_i \qquad (4.2)$$

belongs to M for any vectors $e_i \in L$ and any functions $r_i(\lambda)$ which satisfy the condition

$$\lim_{\lambda \to +0} r_i(\lambda)\lambda^{-1} = 0$$

for all $\lambda > 0$ sufficiently small.

2. For $i \leq 0$,

$$\lim_{\lambda \to +0} \frac{\varphi_i(x(\lambda)) - \varphi_i(x_o)}{\lambda} \leq h_i(e), \qquad (4.3)$$

where $x(\lambda)$ is defined by (4.2) and $h_i(e)$ is a functional which is convex with respect to e.

3. For $i > 0$,

$$\lim_{\lambda \to +0} \frac{\varphi_i(x(\lambda)) - \varphi_i(x_o)}{\lambda} = h_i(e), \qquad (4.4)$$

where $h_i(e)$ is a linear functional.

<u>Theorem</u> 4.1. <u>If</u> x_o <u>is a solution to Problem</u> (4.1) <u>and Conditions</u> 1-3 <u>are satisfied, then there exist numbers</u> λ_i, <u>not all zero, such that</u>

$$\sum_{i=-m}^{k} \lambda_i h_i(e) \geq 0 \text{ for all } e \in K_M, \tag{4.5}$$

and such that $\lambda_i \geq 0$ *for* $i \leq 0$, *and* $\lambda_i \varphi_i(x_o) = 0$ *for* $i < 0$.

Before beginning the proof, let us establish the following lemma.

Lemma 4.1. *Let* $h_1, \ldots, h_k$ *be linearly independent functionals. Then, in order that* x_o *be a solution to Problem* 4.1, *it is necessary that the convex hull of the set*

$$\tilde{K} = \{\xi : \xi \in E^{|I(x_o)|}, \ \xi_i = h_i(e), \ e \in K_M, \ i \in I(x_o)\}$$

and the set

$$P = \{\zeta : \zeta_i < 0 \text{ if } i \in I(x_o) \text{ and } i \leq 0, \ \zeta_i = 0 \text{ for } i > 0\}$$

have an empty intersection, where

$$I(x_o) = \{i : \varphi_i(x) = 0 \text{ for } i < 0 \text{ or } i \geq 0\},$$

and $|I(x_o)|$ *denotes the number of elements in* $I(x_o)$.

Proof of the lemma. Suppose that the lemma is false. Then there exists a vector $\xi^o \in \text{co}\,\tilde{K}$ such that

$$\xi_i^o < 0 \text{ if } i \in I(x_o) \text{ and } i \leq 0,$$

$$\xi_i^o = 0 \text{ for } i > 0.$$

Since $\xi^o \in \text{co}\,\tilde{K}$, there exist elements $\xi^i \in \tilde{K}$ such that

$$\xi^o = \sum_j \lambda_j \xi^j,$$

where $\lambda_j \geq 0$ and $\sum_j \lambda_j = 1$. Further, since $\xi^i \in \widetilde{K}$ for each i, we have that, for some $e^i \in K_M$,

$$\xi^j = h(e^j),$$

where h(e) is a vector with components $h_i(e)$ for $i \in I(x_o)$.

Let $e^o = \sum_j \lambda_j e^j$. Since K_M is a convex cone, $e^o \in K_M$. By virtue of our assumptions,

$$h_i(e^o) \leq \sum_j \lambda_j h_i(e^j) = \sum_j \lambda_j \xi_i^j = \xi_i^o < 0 \text{ if } i \leq 0 \text{ and } i \in I(x_o),$$

$$h_i(e^o) = \sum_j \lambda_j h_i(e^j) = \sum_j \lambda_j \xi_i^j = \xi_i^o = 0 \text{ for } i > 0.$$

Now let us consider the set of equations

$$\psi_i(\lambda, r_1, r_2, \ldots, r_k) \equiv \varphi_i \left(x_o + \lambda e^o + \sum_{j=1}^{k} r_j a^j \right) = 0 \tag{4.6}$$

for $i = 1, \ldots, k$,

where the $a^j \in L$ are chosen such that $h_i(a^j) = \delta_{ij}$, $\delta_{ij} = 0$ if $i \neq j$ and $\delta_{ij} = 1$ for $i = j$. Such a choice for the vectors a^i is always possible, since the h_i are linearly independent.

It is easy to see that

$$\frac{\partial \psi_i}{\partial \lambda} = h_i(e^o) = 0,$$

$$\frac{\partial \psi_i}{\partial r_j} = h_i(a^j) = \delta_{ij},$$

where the derivatives are evaluated at $\lambda = 0$, $r_j = 0$, $j = 1, \ldots, k$. The implicit function theorem which we shall prove at the end of this chapter implies that the system (4.6) has a solution $r_i(\lambda)$, where

$$\lim_{\lambda \to +0} r_i(\lambda)\lambda^{-1} = 0.$$

Now, consider the points

$$x(\lambda) = x_o + \lambda e^o + \sum_{j=1}^{k} r_j(\lambda) a^j.$$

First, $x(\lambda) \in M$ for sufficiently small λ. This follows from the fact that $e^o \in K_M$ and from Condition 1 of the basic hypotheses.

Second,

$$\varphi_i(x(\lambda)) \leq \varphi_i(x_o) + \lambda h_i(e^o) + o(\lambda) \leq \lambda h_i(e^o) + o(\lambda) < 0$$

for all sufficiently small λ, if $i < 0$ and $i \in I(x_o)$, since $h_i(e^o) < 0$. If $i < 0$ and $i \notin I(x_o)$, then $\varphi_i(x(\lambda)) < 0$ for small λ, because $\varphi_i(x_o) < 0$.

Third, for $i = 0$,

$$\varphi_o(x(\lambda)) \leq \varphi_o(x_o) + \lambda h_o(e^o) + o(\lambda) < \varphi_o(x_o),$$

because $h_o(e^o) < 0$.

Finally, by construction of $r_i(\lambda)$,

$$\varphi_i(x(\lambda)) = 0$$

for $i = 1, \ldots, k$.

Therefore, for small λ, $x(\lambda)$ satisfies all of the constraints, and $\varphi_o(x)$ takes on a value at this point which is less than $\varphi_o(x_o)$. This contradicts the assumption that x_o is a solution to Problem (4. 1). |||

Now we shall prove the theorem.

If $h_1, \ldots, h_k$ are linearly independent, then it is sufficient to set $\lambda_i = 0$ for each $i \leq 0$, and, for $i > 0$, to choose λ_i such that

$$\sum_{i=1}^{k} \lambda_i h_i = 0.$$

But if $h_1, \ldots, h_k$ are linearly dependent, then, on the basis of the lemma, the sets co $\widetilde{K}$ and P have no points in common. Since these are convex sets in a finite-dimensional space, they can be separated. This means that there exists a vector Λ, with components λ_i where $i \in I(x_o)$, such that

$$\sum_{i \in I(x_o)} \lambda_i \xi_i \geq 0 \geq \sum_{i \in I(x_o)} \lambda_i \zeta_i,$$

for all $\xi \in \widetilde{K}$ and $\zeta \in P$. Since every $\xi \in \widetilde{K}$ can be represented in the form

$$\xi = h(e), \quad e \in K_M,$$

we obtain $\sum_{i \in I(x_o)} \lambda_i h_i(e) \geq 0$ for all $e \in K_M$. It is easy to see that the right-hand side of Inequality (4. 7) implies that $\lambda_i \geq 0$ for all $i \in I(x_o)$ with $i \leq 0$. But $\varphi_i(x_o) = 0$ if $i \in I(x_o)$ and $i < 0$. Thus, $\lambda_i \varphi_i(x_o) = 0$ if $i < 0$ and $i \in I(x_o)$.

If we now set $\lambda_i = 0$ for $i \notin I(x_o)$, we obtain all of the desired conclusions of the theorem. |||

<u>Corollary</u>. If there are no equality constraints, or if $\varphi_i(x)$ is linear for each $i > 0$, then, in the basic assumptions 1-3, we can set

$$x(\lambda) = x_o + \lambda e, \quad r_i \equiv 0, \quad i > 0.$$

Indeed, in this case, $\varphi_i = h_i$, and the system (4. 6) takes the form

$$\varphi_i\left(x_o + \lambda e^o + \sum_{j=1}^{k} r_j a^j\right) = \varphi_i(x_o) + \lambda h_i(e^o) + \sum_{j=0}^{k} r_j h_j(a^j) = 0.$$

But, since $\varphi_i(x_o) = 0$ by assumption and $h_i(e^o) = 0$ by construction of e^o, to satisfy the thus obtained system of equations, it is sufficient to set $r_j \equiv 0$ for $j = 1, \ldots, k$. |||

We shall now particularize Theorem 4. 1 to the case where the space under consideration is a Banach space, and the functionals $\varphi_i(x)$ for $i \leq 0$ are quasi-differentiable. By definition of a quasi-differentiable functional,

$$\lim_{\lambda \to +0} \frac{\varphi_i(x_o + \lambda e) - \varphi_i(x_o)}{\lambda} = \sup_{x^* \in M_i(x_o)} x^*(e).$$

If we also suppose that the $\varphi_i(x)$ satisfy a Lipschitz condition, then it is easy to verify that formula (4.3) holds with

$$h_i(e) = \sup_{x^* \in M_i(x_o)} x^*(e),$$

and, in addition, that the inequality in (4.3) must be replaced by an equality.

Now let $\varphi_i(x)$ for $i > 0$ be such that (4.4) holds, where h_i coincides with some functional $x_i^* \in B^*$. Then all of the hypotheses of Theorem 4.1 are satisfied, and, therefore, if x_o is a solution to Problem 4.1, then there exist numbers λ_i such that

$$\sum_{i=-m}^{0} \lambda_i \sup_{x_i^* \in M_i(x_o)} x_i^*(e) + \sum_{i=1}^{k} \lambda_i x_i^*(e) \geq 0 \qquad (4.8)$$

for all $e \in K_M$.

<u>Lemma</u> 4.2. <u>If the sets</u> $M_i(x_o)$ <u>are bounded, then, for Condition</u> (4.8) <u>to be satisfied, it is necessary and sufficient that there exist functionals</u> $x_i^* \in M_i(x_o)$ <u>such that</u>

$$\sum_{i=-m}^{k} \lambda_i x_i^* \in K_M^* . \tag{4.9}$$

Proof. Consider the set $N^* = \left\{x^*: x^* = \sum_{i=-m}^{k} \lambda_i x_i^*, x_i^* \in M_i(x_o), i \leq 0\right\}$. Obviously, this set is convex, since the sets $M_i(x_o)$ are. Moreover, it is weak* closed and weak* compact since, by assumption (see Definition 3.1), the $M_i(x_o)$ are weak* closed and bounded and, thus, weak* compact. Let us suppose that (4.9) does not hold, i.e., that N^* and K_M^* have an empty intersection. Then $K_M^* - N^*$ is regularly convex, because K_M^* is weak* closed, N^* is weak* closed and compact, and, thus, $K_M^* - N^*$ is also weak* closed. Since K_M^* and N^* have no points in common, $K_M^* - N^*$ does not contain the zero functional. Therefore, there exists an $e \in B$ such that

$$y^*(e) - x^*(e) \geq \delta > 0 \tag{4.10}$$

for all $y^* \in K_M^*$ and $x^* \in N^*$. The inequality which has just been obtained shows that $y^*(e)$ is bounded for all $y^* \in K_M^*$. But, since K_M^* is a cone,

$$\inf_{y^* \in K_M^*} y^*(e) = 0 . \tag{4.11}$$

This implies that $e \in \overline{K}_M$.

Furthermore, a comparison of (4.10) and (4.11)

shows that $x^*(e) \leq -\delta$ for all $x^* \in N^*$.

But $e \in \overline{K}_M$ and (4.8) imply that there exists an $x^* \in N^*$ such that

$$x^*(e) \geq -\frac{\delta}{2}, \quad x^* \in N^*.$$

Thus, we have obtained a contradiction. This means that N^* and K^*_M have a non-empty intersection and, therefore, (4.9) holds. The sufficiency is obvious. |||

Theorem 4.2. Let $\varphi_i(x)$, for $i = -m, \ldots, 0, \ldots, k$, be functionals on a Banach space. If $\varphi_i(x)$ satisfies a Lipschitz condition and is quasi-differentiable for each $i \leq 0$, and satisfies a Lipschitz condition and has a Gâteaux differential x^*_i for each $i > 0$, then, in order that x_o be a solution to Problem (4.1), it is necessary that there exist numbers λ_i (not all zero) and functionals $x^*_i \in M_i(x_o)$, for $i \leq 0$, such that

$$\sum_{i=-m}^{k} \lambda_i x^*_i \in K^*_M,$$

$$\lambda_i \geq 0 \text{ for } i \leq 0 \text{ and } \lambda_i \varphi_i(x_o) = 0 \text{ for } i \neq 0.$$

The theorem actually follows from the preceding arguments. It is only necessary to prove some details.

First, if, for each $i > 0$, $\varphi_i(x)$ satisfies a Lipschitz condition, then

$$\lim_{\lambda \to +0} \frac{\varphi_i(x(\lambda)) - \varphi_i(x_o)}{\lambda} = x_i^*(e),$$

where $x(\lambda)$ is given by formula (4. 2).

The proof of this fact is trivial. From this it follows that Condition 3 of the basic assumptions is satisfied.

Second, let us show that $M_i(x_o)$ is bounded, as is required in Lemma 4. 2. If we assume the contrary, then, for every integer $n > 0$, there exist an element e_n with $\|e_n\| = 1$ and a functional $x_n^* \in M_i(x_o)$ such that

$$x_n^*(e_n) \geq n - \varepsilon,$$

for some $\varepsilon > 0$. Thus,

$$\varphi_i(x_o + \lambda e_n) - \varphi_i(x_o)$$

$$= \lambda \sup_{x^* \in M_i(x_o)} x^*(e_n) + o(\lambda) \geq \lambda(n - \varepsilon) + o(\lambda).$$

On the other hand, by virtue of the Lipschitz condition,

$$|\varphi_i(x_o + \lambda e_n) - \varphi_i(x_o)| \leq L\lambda.$$

Therefore,

$$L\lambda \geq \lambda(n - \varepsilon) + o(\lambda).$$

But if $n - \varepsilon > L$, then the last inequality cannot hold for sufficiently small λ. The contradiction which we have just obtained shows that $M_i(x_o)$ is bounded. This completes the proof of the theorem. |||

Theorem 4.3 (Kuhn-Tucker). Let $\varphi_i(x)$ be convex bounded functionals on B, and, moreover, suppose that $\varphi_i(x)$ is linear for each $i > 0$. Also suppose that the set M is convex. Then, in order that x_o be a solution to the minimization Problem (4.1), it is necessary that there exist numbers λ_i, with $\lambda_i \geq 0$ for $i \leq 0$, such that

$$\sum_{i=-m}^{k} \lambda_i \varphi_i(x) \geq \sum_{i=-m}^{k} \lambda_i \varphi_i(x_o) \text{ for all } x \in M. \tag{4.12}$$

Moreover,

$$\lambda_i \varphi_i(x_o) = 0 \text{ for } i \neq 0. \tag{4.13}$$

If $\lambda_o > 0$, then the conditions of the theorem are sufficient.

Proof. We define the cone K_M as follows:

$$K_M = \{e : e = \lambda(x - x_o) \text{ for } x \in M,\ x \neq x_o,\ \lambda > 0\}.$$

It is easy to see that if $e \in K_M$, then

$$x(\lambda) = x_o + \lambda e \in M$$

for small λ. Further,

$$\lim_{\lambda \to +0} \frac{\varphi_i(x_o + \lambda e) - \varphi(x_o)}{\lambda} = \frac{\partial \varphi(x_o)}{\partial e}$$

$$\leq \varphi_i(x_o + e) - \varphi(x_o) \equiv h_i(e).$$

For $i > 0$, the inequality in an obvious way becomes an equality. For $i \leq 0$, this inequality follows from the fact that the quotient

$$\frac{\varphi_i(x_o + \lambda e) - \varphi(x_o)}{\lambda}$$

is monotonically decreasing for decreasing λ (see Lemma 6 of the introduction). If we take into account the corollary to Theorem 4.1, we can see that all of the hypotheses of Theorem 4.1 are satisfied. Thus, there exist numbers λ_i, with $\lambda_i \geq 0$ for $i \leq 0$, such that

$$\sum_{i=-m}^{k} \lambda_i h_i(e) = \sum_{i=-m}^{k} \lambda_i \left[\varphi_i(x_o + e) - \varphi(x_o)\right] \geq 0$$

for all $e \in K_M$.

Setting here $e = x - x_o$, where $x \in M$, we obtain

$$\sum_{i=-m}^{k} \lambda_i \varphi_i(x) \geq \sum_{i=-m}^{k} \lambda_i \varphi_i(x_o)$$

for all $x \in M$.

Now suppose that (4.12) and (4.13) are satisfied and that $\lambda_o > 0$. Then x_o is a solution to Problem (4.1). Indeed, for any x which satisfies all of the constraints, we have, by virtue of (4.12) and (4.13),

$$\lambda_o \varphi_o(x_o) = \sum_{i=-m}^{k} \lambda_i \varphi_i(x_o) \leq \sum_{i=-m}^{k} \lambda_i \varphi_i(x) \leq \lambda_o \varphi_o(x)$$

because $\varphi_i(x) \leq 0$ for each $i < 0$ and $\varphi_i(x) = 0$ for each $i > 0$.

Thus,

$$\varphi_o(x_o) \leq \varphi_o(x),$$

which implies that x_o is a solution to Problem (4. 1). |||

Let us now establish the relationship between the results which have been obtained and the theory of Dubovitskii and Milyutin [10].

Suppose that a functional $\varphi(x)$ is given on B, and that we wish to minimize φ on L subject to the constraints $x \in \Omega_i$, $i = 1, \ldots, n$. Let x_o be a minimum point. Assume that, at x_o, there exists, for every Ω_i, a convex cone K_i such that, whenever $e \in K_i$,

$$x(\lambda) = x_o + \lambda e' \in \Omega_i \tag{4. 14}$$

for all sufficiently small $\lambda > 0$ and all e' which satisfy the inequality $\|e' - e\| \leq \varepsilon_e$, where $\varepsilon_e > 0$ depends on e. Moreover, suppose that there exists a subspace Z tangent to L at x_o, i. e., for every $e \in L$ there exists a $r(\lambda)$ such that

$$x(\lambda) = x_o + \lambda e + r(\lambda) \in L \tag{4. 15}$$

whenever λ is sufficiently small. Here, $\|r(\lambda)\| \lambda^{-1} \to 0$ as

$\lambda \to +0$.

Further, suppose that $\varphi(x)$ is such that there exists a cone of "forbidden variations", i.e., a convex cone K_o such that (4.14) holds for Ω_o, where

$$\Omega_o = \{x: \varphi(x) < \varphi(x_o)\}.$$

Now the first theorem of Milyutin and Dubovitskii is almost obvious.

Theorem 4.4. *In order that x_o be a solution to the problem*

$$\min \varphi(x),$$

$$x \in \Omega_i \text{ for } i = 1, \ldots, n,$$

$$x \in L$$

it is necessary that the cones $K_o, K_1, \ldots, K_n$ and Z have an empty intersection.

Proof. Suppose that $K_o, K_1, \ldots, K_n$ and Z have a non-empty intersection, i.e., that there exists an e such that $e \in Z$ and $e \in K_i$ for each i. By definition of Z, there exists a function $r(\lambda)$ such that

$$x = x_o + \lambda e + r(\lambda) \in L$$

and

$$\lim_{\lambda \to 0} \frac{\|r(\lambda)\|}{\lambda} = 0.$$

Further, since $e \in K_i$,

$$x = x_o + \lambda e' \in \Omega_i$$

whenever λ is sufficiently small, and

$$\| e' - e \| \leq \varepsilon_e .$$

Let us set

$$e' = e + \frac{r(\lambda)}{\lambda} .$$

But

$$\| e' - e \| = \frac{\| r(\lambda) \|}{\lambda} ,$$

and this quotient tends to zero. Thus, for sufficiently small λ, $\| e' - e \|$ is less than ε_e, and, therefore,

$$x = x_o + \lambda e + r(\lambda) \in \Omega_i .$$

Thus, for sufficiently small λ, $x \in L$ and $x \in \Omega_i$ for $i = 0, 1, \ldots, n$. This means that x satisfies all of the constraints and, moreover,

$$\varphi(x) < \varphi(x_o) ,$$

since $x \in \Omega_o$. The last inequality contradicts the fact that x_o yields a minimum for $\varphi(x)$. |||

In order to employ Theorem (4.4), it is necessary to give an effective condition for the cones to have an empty intersection. An answer to this problem is provided by the second theorem of Dubovitskii and Milyutin.

Theorem 4.5. *In order that the cones $K_0, K_1, \ldots, K_n$ and Z have an empty intersection, it is necessary and sufficient that there exist functionals $x_i^* \in K_i^*$, $i = 0, 1, \ldots, n$, and $x^* \in Z^*$ such that*

$$x_0^* + x_1^* + \ldots + x_n^* + x^* = 0 \tag{4.16}$$

(K_i^* is the dual cone to K_i), where the x_i^* and x^* are not all zero.

Proof. There is a very elegant proof of this theorem due to its authors. Here we shall state a proof which is based upon the preceding results. We notice at once that the cones K_i, $i=0,\ldots,n$, are open by construction. Consider their intersection. Without loss of generality, we shall assume that it is non-empty. Thus, by the assumption of the theorem,

$$\left(\bigcap_{i=0}^{n} K_i\right) \cap Z = \emptyset .$$

But $\bigcap_{i=0}^{n} K_i$ is an open, convex cone, and Z is a convex set. By virtue of our separation theorem, there exists a functional $y_o^* \in B^*$ such that

$$\left.\begin{aligned} y_o^*(e) &\geq 0 \quad \text{for} \quad e \in \bigcap_{i=0}^{n} K_i, \\ y_o^*(e) &\leq 0 \quad \text{for} \quad e \in Z . \end{aligned}\right\} \tag{4.17}$$

We set

$$\mu_0(x) = y_0^*(x) ,$$

$$\mu_i(x) = \max_{x^* \in M_i} x^*(x) \quad \text{for } i = 0, 1, \ldots, n ,$$

where $M_i = (-K_i^*) \cap S^*$, and S^* is the unit ball in B^*, i.e., $S^* = \{ x^*: \|x^*\| \leq 1\}$. It is not difficult to see that the $\{x: \mu_i(x) \leq 0\} = \overline{K}_i$.

Indeed, this follows from the fact that $x \in \overline{K}_i$ if and only if

$$x^*(x) \geq 0 \quad \text{for all} \quad x^* \in K_i^* .$$

Now it is clear that the set

$$\Omega = \{ x: \mu_i(x) \leq 0 \quad \text{for} \quad i = 0, \ldots, n\}$$

coincides with $\bigcap_{i=0}^{n} \overline{K}_i$. Inequalities (4.17) imply that the functional $\mu_0(x)$ achieves its minimum on Ω at $x_0 = 0$.

We now note the following. First, since all of the K_i are open, their intersection is also open, and there exists an x_1 which is an interior point of $\bigcap_{i=0}^{n} K_i$.

But then $\mu_i(x_1) < 0$ for each $i = 0, \ldots, n$.

Second, by virtue of Theorem 1.6 of Chapter I, for each $i > 0$, the set $M_i(0)$ of support functionals to $\mu_i(x)$ at $x_0 = 0$ coincides with M_i.

On the basis of Theorem 2.4, we can now assert that there exist numbers $\lambda_i \leq 0$ such that

$$y_0^* = \sum_{i=0}^{n} \lambda_i y_i^* ,$$

where $y_i^* \in M_i$. Now, set $x^* = -y_0^*$ and $x_i^* = \lambda_i y_i^*$. Then the preceding equation is transformed into

$$x_0^* + x_1^* + \ldots + x_n^* + x^* = 0 .$$

But $x^* \in Z^*$ by virtue of (4.17). Further, $x_i^* \in K_i^*$, since $x_i^* = \lambda_i y_i^*$, $\lambda_i \leq 0$, and $y_i^* \in (-K_i^*) \cap S^* \subset -K_i^*$.

The necessity of the conditions of the theorem has been proved.

The sufficiency of Condition (4.16) is easy to prove by contradiction.

Indeed, suppose that (4.16) is satisfied, and suppose that there exists an e_0 such that $e_0 \in K_i$ for $i = 0, \ldots, n$, and such that $e_0 \in Z$. Then

$$x_i^*(e_0) \geq 0 \text{ for each } i \text{ and } x^*(e_0) \geq 0$$

by definition of the dual cone, and at least one number $x_i^*(e_0)$ is positive, because, for each $i = 0, \ldots, n$, $e_0 \in K_i$ and K_i is open (and, thus, e_0 is an interior point of K_i), and, moreover, not all of the x_i^* are zero.

This implies that

$$x_0^*(e) + x_1^*(e_0) + \ldots + x_n^*(e_0) + x^*(e_0) > 0,$$

which contradicts (4.16). |||

Thus, we see that the preceding results imply the theorem of Dubovitskii and Milyutin. This theorem yields very general conditions for an extremum. In particular, it is in some respects more general than Theorem 4.1, since it does not require the equality constraints (whose role is played by the set L) to be given in the form of a finite system of functionals which must equal zero. On the other hand, Theorems 4.4 and 4.5 by themselves do not permit us to construct necessary conditions in the general case, because they leave open the question of how to effectively construct the cones and the subspace Z.

Our method of presentation in this work has been as follows: First, we developed ways of evaluating directional derivatives, or equivalently, of constructing the cones K_i. Then we formulated the necessary conditions. Because of our success in the first task, and within the limits thereof, we could immediately write down these conditions in concrete form.

We shall now consider a generalization of the problem studied at the beginning of this chapter. This generalization permits us to obtain, for some problems (e.g., the discrete optimal control problem), conditions for a minimum which are finer than those which are provided by Theorem 4.1.

Let L be a linear space, let X be a subset of L, and let U be some arbitrary set. Let there be given functionals $\varphi_i(x, \alpha)$ for $i = -m, \ldots, -1, 0, 1, \ldots, k$, on $X \times U$. We are seeking necessary conditions for the points x_0, u_0 to be a solution to the problem

$$\left.\begin{aligned} &\min \varphi_0(x, u) \,, \\ &\varphi_i(x, u) \le 0 \quad \text{for} \quad i < 0 \,, \\ &\varphi_i(x, u) = 0 \quad \text{for} \quad i > 0 \,, \\ &x \in X \,, \quad u \in U \,. \end{aligned}\right\} \tag{4.18}$$

We shall state some basic assumptions under which the problem will be solved.

1. If $\varphi(x, u) = (\varphi_{-m}(x, u), \ldots, \varphi_0(x, u), \ldots, \varphi_k(x, u))$, then for every fixed x, the set $\varphi(x, U) = \{\varphi(x, u) : u \in U\}$ is convex.

2. There exists at x_0 a convex cone K_X such that $e \in K_X$ implies that

$$x(\lambda) = x_0 + \lambda e + \sum_{j=1}^{k} r_j(\lambda) a_j \in X \tag{4.19}$$

for sufficiently small $\lambda > 0$ and for any $a_j \in L$, and for any functions $r_j(\lambda)$ which satisfy the condition

$$\lim_{\lambda \to +0} \frac{r_j(\lambda)}{\lambda} = 0 .$$

3. For $i \leq 0$ and for any $u \in U$,

$$\lim_{\lambda \to +0} \frac{\varphi_i(x(\lambda), u) - \varphi_i(x_0, u)}{\lambda} \leq h_i(u, e) ,$$

where $x(\lambda)$ is given by (4.19), and $h_i(u, e)$ is a functional which is convex in e and is such that $h_i(u, \lambda e) = \lambda h_i(u, e)$ for every e and $\lambda > 0$.

4. For each $i > 0$, there exists a functional $h_i(u, e)$ which is linear in e and is such that

$$\varphi_i(x(\lambda), u) - \varphi_i(x_0, u) - \lambda h_i(u, e) - \sum_{j=1}^{k} r_j h_i(u, a_j)$$
$$= r\left(\sqrt{\lambda^2 + \sum_{j=1}^{k} r_j^2}\right),$$

where $x(\lambda)$ is given by

$$x(\lambda) = x_0 + \lambda e + \sum_{j=1}^{k} r_j a_j , \tag{4.20}$$

and $r(z)$ is a function such that

$$\lim_{z \to 0} \frac{r(z)}{z} = 0.$$

Moreover, $\varphi_i\left(x_0 + \lambda e + \sum_{j=1}^{k} r_j a_j, u\right)$ is, for each i, a continuous function of λ and r_j.

Condition 4 can be interpreted in a different way as follows. If we substitute Expression (4.20) into $\varphi(x,u)$, then we obtain a function of a finite number of variables: λ and r_j. Condition 4 then means that this function is differentiable at x_0, i.e., this function can be approximated in some neighborhood of x_0 by a linear function of λ and r_j, where the error consists of terms of higher order than λ and r_j.

Let us now discuss what Condition 1 implies. Indeed, it follows from this condition that, if $u_j \in U$ for each $j=1,\ldots,n$, and if the numbers γ_j satisfy the condition

$$\sum_{j=1}^{n} \gamma_j = 1, \quad \gamma_j \geq 0 \text{ for each } j,$$

then there exists a $\bar{u} \in U$ such that

$$\varphi(x,\bar{u}) = \sum_{j=1}^{n} \gamma_j \varphi(x,u_j) . \tag{4.21}$$

Further, if $u_0 \in U$ and $\bar{u} \in U$, then there exists a $u(x,\lambda)$ such that

$$\varphi(x,u(x,\lambda)) = \varphi(x,u_0) + \lambda[\varphi(x,\bar{u}) - \varphi(x,u_0)] \tag{4.22}$$

for $0 \le \lambda \le 1$.

Now, let x_0, u_0 be a solution to Problem (4.18), and suppose that the functionals $h_i(u_0, e)$, $i = 1, \ldots, k$, are linearly independent. Moreover, without loss of generality, we can assume that $\varphi_i(x_0, u_0) = 0$ for each $i < 0$.

Consider the sets $\tilde{K}$ and P defined by

$$\tilde{K} = \{ \xi : \xi = h(u_0, e) + [\varphi(x_0, u) - \varphi(x_0, u_0)],\ e \in K_X, u \in U \},$$

where ξ is an $(m+k+1)$-dimensional vector and $h(u_0, e) = (h_{-m}(u_0, e), \ldots, h_k(u_0, e))$, and

$$P = \{ \zeta : \zeta_i < 0 \text{ for } i \le 0 \text{ and } \zeta_i = 0 \text{ for } i > 0 \}.$$

We shall show that P and the convex hull of $\tilde{K}$ have an empty intersection.

Assume the contrary. Then there exists a vector $\xi^0 \in \mathrm{co}\,\tilde{K}$ such that

$$\left.\begin{aligned} \xi_i^0 &< 0 \quad \text{for } i \le 0 , \\ \xi_i^0 &= 0 \quad \text{for } i > 0 , \end{aligned}\right\} \tag{4.23}$$

By definition of a convex hull and by definition of $\tilde{K}$, there exist non-negative numbers γ_j for $j = 1, \ldots, n$, with $\sum_{j=1}^{n} \gamma_j = 1$, and elements $e_j \in K_X$ and $u_j \in U$ such that

$$\xi^0 = \sum_{j=1}^{n} \gamma_j \left[h(u_0, e_j) + [\varphi(x_0, u_j) - \varphi(x_0, u_0)] \right].$$

Now set

$$e_0 = \sum_{j=1}^{n} \gamma_j e_j ,$$

and define $\bar{u}$ by the condition

$$\varphi(x_0, \bar{u}) = \sum_{j=1}^{n} \gamma_j \varphi(x_0, u_j) .$$

Since K_X is a convex cone and $e_j \in K_X$ for all j, $e_0 \in K_X$. It follows from the convexity of $h_i(u_0, e)$ in e for each $i \leq 0$, the linearity of these functionals for each $i > 0$, and (4.23) that

$$\left.\begin{array}{l} h_i(u_0, e_0) + [\varphi_i(x_0, \bar{u}) - \varphi_i(x_0, u_0)] \leq \xi_i^0 < 0 \text{ for } i \leq 0, \\ h_i(u_0, e_0) + [\varphi_i(x_0, \bar{u}) - \varphi_i(x_0, u_0)] = \xi_i^0 = 0 \text{ for } i > 0. \end{array}\right\} \quad (4.24)$$

Since the $h_i(u_0, e)$, $i = 1, \ldots, k$, are linearly independent, there exist elements $a_j \in L$ such that

$$h_i(u_0, a_j) = \delta_{ij} \text{ for } i, j = 1, \ldots, k,$$

where $\delta_{ij} = 0$ if $i \neq j$ and $\delta_{ij} = 1$ for $i = j$.

Consider the system of nonlinear equations

$$\psi_i(\lambda, r_1, \ldots, r_k) = 0 \text{ for } i = 1, \ldots, k, \quad (4.25)$$

where

$$\psi_i(\lambda, r_1, \ldots, r_k) = \varphi_i\left(x_0 + \lambda e_0 + \sum_{j=1}^{k} r_j a_j, u(\lambda, r_1, \ldots, r_k)\right),$$

and $u(\lambda, r_1, \ldots, r_k)$ with $0 \leq \lambda \leq 1$ is defined by the condition

$$\varphi_i\left(x_0 + \lambda e_0 + \sum_{j=1}^{k} r_j a_j, u(\lambda, r_1, \ldots, r_k)\right)$$

$$= \varphi_i\left(x_0 + \lambda e_0 + \sum_{j=1}^{k} r_j a_j, u_0\right) + \lambda\left[\varphi_i\left(x_0 + \lambda e_0 + \sum_{j=1}^{k} r_j a_j, \bar{u}\right)\right.$$

$$\left. - \varphi_i\left(x_0 + \lambda e_0 + \sum_{j=1}^{k} r_j a_j, u_0\right)\right] \text{ for } i = -m, \ldots, k. \tag{4.26}$$

Thus, $\psi_i(\lambda, r_1, \ldots, r_k)$ coincides with the right-hand side of (4.26). If we now make use of Condition 4 of the hypotheses under which we are solving the problem, and separate out the linear part of the right-hand side of (4.26), we obtain

$$\psi_i(\lambda, r_1, \ldots, r_k) = \lambda h_i(u_0, e_0) + \sum_{j=1}^{k} r_j h_i(u_0, a_j)$$

$$+ \lambda[\varphi_i(x_0, \bar{u}) - \varphi_i(x_0, u_0)] + \ldots = \sum_{j=1}^{k} r_j h_i(u_0, a_j) + \ldots,$$

where the dots denote terms which may be estimated by the term

$$r\left(\sqrt{\lambda^2 + \sum_{j=1}^{k} r_j^2}\right),$$

and we have made use of the second of Relations (4.24).

On the basis of Theorem 4.7, which is stated at the end of this chapter, we can assert that system (4.25) can be solved for $r_1, \ldots, r_k$; i.e., for sufficiently small λ, the system determines function $r_j(\lambda)$ which, moreover, satisfy the condition

$$\lim_{\lambda \to 0} \frac{r_j(\lambda)}{\lambda} = 0.$$

Now set

$$x(\lambda) = x_0 + \lambda e_0 + \sum_{j=1}^{n} r_j(\lambda) a_j ,$$

$$u(\lambda) = u(\lambda, r_1(\lambda), \ldots, r_k(\lambda)).$$

Then, by definition of the functions ψ_i and $u(\lambda, r_1, \ldots, r_k)$, we obtain

$$\varphi_i(x(\lambda), u(\lambda)) = 0 \quad \text{for } i=1, \ldots, k .$$

Further, on the basis of Assumption 2, $x(\lambda) \in X$ for sufficiently small $\lambda > 0$. Moreover, for $i \leq 0$,

$$\varphi_i(x(\lambda), u(\lambda)) = \varphi_i(x(\lambda), u_0) + \lambda[\varphi_i(x(\lambda), \bar{u}) - \varphi_i(x(\lambda), u_0)],$$

$$\lim_{\lambda \to +0} \frac{\varphi(x(\lambda), u(\lambda)) - \varphi_i(x_0, u_0)}{\lambda}$$

$$= \lim_{\lambda \to +0} \frac{\varphi_i(x(\lambda), u_0) - \varphi_i(x_0, u_0)}{\lambda}$$

$$+ \lim_{\lambda \to +0} [\varphi_i(x(\lambda), \bar{u}) - \varphi_i(x(\lambda), u_0)]$$

$$\leq h_i(u_0, e_0) + [\varphi_i(x_0, \bar{u}) - \varphi_i(x_0, u_0)] .$$

But, (4.24) implies that the right-hand side of the last formula is negative. Therefore,

$$\varphi_i(x(\lambda), u(\lambda)) < \varphi_i(x_0, u_0) \quad \text{for } i \leq 0.$$

The just obtained results contradict the assumption that x_0, u_0 is a solution to Problem (4.18). Indeed, a point $x(\lambda), u(\lambda)$ satisfies all of the constraints, because $x(\lambda) \in X$ for sufficiently small λ, $u(\lambda) \in U$ by definition,

$$\varphi_i(x(\lambda), u(\lambda)) = 0 \quad \text{for } i > 0,$$

$$\varphi_i(x(\lambda), u(\lambda)) < \varphi_i(x_0, u_0) = 0 \text{ for } i < 0,$$

and $\varphi_0(x(\lambda), u(\lambda)) < \varphi_0(x_0, u_0)$, i.e., x_0, u_0 does not yield the minimum value for $\varphi_0(x, u)$, subject to the given constraints.

The contradiction which we have just obtained shows that co $\widetilde{K}$ and P have an empty intersection. But co $\widetilde{K}$ and P are convex sets in an (m+k+1)-dimensional space. Therefore, there exist constants λ_i, i= -m,...,k, not all zero, such that

$$\sum_{i=-m}^{k} \lambda_i \xi_i \geq 0 \geq \sum_{i=-m}^{k} \lambda_i \zeta_i \quad \text{for all } \xi \in \text{co}\,\widetilde{K} \text{ and } \zeta \in P.$$

The right-hand side of this inequality implies that $\lambda_i \geq 0$ for each $i \leq 0$. The left-hand side yields, by definition, that

$$\sum_{i=-m}^{k} \lambda_i [\, h_i(u_0, e) + [\, \varphi_i(x_0, u) - \varphi_i(x_0, u_0)]\,] \geq 0$$

for all $e \in K_X$ and $u \in U$.

But since e and u vary independently, the preceding inequality is equivalent to the following two relations:

$$\left.\begin{aligned} &\sum_{i=-m}^{k} \lambda_i h_i(u_0, e) \geq 0 \quad \text{for all} \quad e \in K_X, \\ &\sum_{i=-m}^{k} \lambda_i \varphi_i(x_0, u_0) = \min_{u \in U} \sum_{i=-m}^{k} \lambda_i \varphi_i(x_0, u). \end{aligned}\right\} \qquad (4.27)$$

Thus we have proved the following theorem.

Theorem 4.6. Suppose that Assumptions 1-4 are satisfied and that the functionals $h_i(u_0, e)$, $i = 1, \ldots, k$, are linearly independent. Then, in order that x_0, u_0 be a solution to Problem (4.18), it is necessary that there exist numbers λ_i, $i = -m, \ldots, k$, not all zero, such that Relations (4.27) are satisfied and such that

$$\lambda_i \geq 0 \text{ for } i \leq 0 \text{ and } \lambda_i \varphi_i(x_0, u_0) = 0 \text{ for } i < 0. \qquad (4.28)$$

Of the assertions of the theorem, an explanation is required only for the equation

$$\lambda_i \varphi_i(x_0, u_0) = 0 \quad \text{for} \quad i < 0 .$$

Under the assumptions which we made in order to prove the

theorem, i.e., $\varphi_i(x_0, u_0) = 0$ for each $i < 0$, this equation is obvious. But if, for some i,

$$\varphi_i(x_0, u_0) < 0 ,$$

then such constraints can be ignored while constructing $x(\lambda)$ and $u(\lambda)$, since, for such an i,

$$\begin{aligned}\varphi_i(x(\lambda), u(\lambda)) &= \varphi_i(x(\lambda), u_0) \\ &+ \lambda[\varphi_i(x(\lambda), \bar{u}) - \varphi_i(x(\lambda), u_0)] \leq \varphi_i(x_0, u_0) \\ &+ \lambda[h_i(x_0, e_0) + [\varphi_i(x_0, \bar{u}) - \varphi(x_0, u_0)]] + o(\lambda).\end{aligned}$$

The inequality in the last relation follows from Assumption 3, and the term to the right of the inequality sign becomes negative for sufficiently small λ, independently of the choice of e_0 and $\bar{u}$. Therefore, if $\varphi_i(x_0, u_0) < 0$ for some $i < 0$, we can repeat the entire proof of the theorem as if there were no such constraints. In addition, in the derivation of the formulas (4.27), one ought to set $\lambda_i = 0$ for all such i, which automatically implies that Conditions (4.28) are satisfied.

The hypotheses which we imposed on the equality-type constraints in Theorems 4.1 and 4.6 (i.e., Condition 3 of Theorem 4.1 and Condition 4 of Theorem 4.6) were of different form. Let us show that these conditions are in actuality equivalent.

Indeed, if Condition 4 of Theorem 4.6 is satisfied, then, for

$$x(\lambda) = x_0 + \lambda e + \sum_{j=1}^{k} r_j(\lambda) a_j, \tag{4.29}$$

$$\frac{\varphi_i(x(\lambda), u) - \varphi_i(x_0, u)}{\lambda} = h_i(u, e) + \sum_{j=1}^{k} \frac{r_j(\lambda)}{\lambda} h_i(u, a_j)$$

$$+ \frac{1}{\lambda} r\left(\sqrt{\lambda^2 + \sum_{j=1}^{n} r_j^2(\lambda)}\right) \to h_i(u, e), \tag{4.30}$$

if $r_j(\lambda)\lambda^{-1} \to 0$ as $\lambda \to 0$. Thus, if Condition 4 of Theorem 4.6 holds, then Condition 3 of Theorem 4.1 also holds.

Conversely, suppose that Condition 3 is satisfied. We shall show that Condition 4 of Theorem 4.6 is then also satisfied. In order to prove this, we assume the contrary, i.e., we assume that there exist sequences $\{\lambda_n\}$ and $\{r_1^n\}, \ldots, \{r_k^n\}$ such that $\lambda_n \to 0$ and $r_j^n \to 0$ as $n \to \infty$. But

$$\varepsilon_n = \frac{1}{\sqrt{\lambda_n^2 + \sum_{j=1}^{k} (r_j^n)^2}} \left[\varphi_i\left(x_0 + \lambda_n e + \sum_{j=1}^{k} r_j^n a_j, u\right)\right.$$

$$\left. - \varphi_i(x_0, u) - \lambda_n h_i(u, e_n) - \sum_{j=1}^{k} r_j^n h_i(u, a_j)\right]$$

does not tend to zero, $|\varepsilon_n| \geq \delta$ for all n. Let

$$\lambda'_n = \sqrt{\lambda_n^2 + \sum_{j=1}^{k} (r_j^n)^2}\,, \quad \gamma_n = \frac{\lambda_n}{\lambda'_n}\,, \quad \omega_j^n = \frac{r_j^n}{\lambda'_n}\,.$$

Since, by definition,

$$\sqrt{\gamma_n^2 + \sum_{j=1}^{n} (\omega_j^n)^2} = 1\,,$$

we can choose a convergent subsequence of the sequence of vectors $\{\gamma_n, \omega_1^n, \ldots, \omega_k^n\}$. Thus, we shall assume that

$$\begin{aligned} \gamma_n &\to \gamma_0\,, \\ \omega_j^n &\to \omega_j^0\,, \end{aligned} \tag{4.31}$$

and set

$$e' = \gamma_0 e + \sum_{j=1}^{k} \omega_j^0 a_j\,.$$

Now ε_n can be written in the following form:

$$\begin{aligned} \varepsilon_n = &\Big\{\varphi_i\Big(x_0 + \lambda'_n e' + \lambda'_n(\gamma_n - \gamma_0)e \\ &+ \sum_{j=1}^{k} \lambda'_n(\omega_j^n - \omega_j^0)a_j, u\Big) - \varphi_i(x_0, u)\Big\}(\lambda'_n)^{-1} \\ &- h_i(ue') + (\gamma_n - \gamma_0)h_i(u, e) + \sum_{j=1}^{k} (\omega_j^n - \omega_j^0)h_i(u, a_j)\,. \end{aligned}$$

Condition 3 of Theorem 4.1 and (4.31) imply that, as $n \to \infty$, i.e., as $\lambda'_n \to 0$, the first term of the formula for ε_n tends to $h_i(u, e')$, and ε_n itself tends to zero. But this contradicts the assumption that $|\varepsilon_n| \geq \delta$. This proves that Condition 3 of Theorem 4.1 implies Condition 4 of Theorem 4.6.

In concluding this chapter, let us give a proof of the theorem on the solvability of a system of non-linear equations, which has been frequently referred to.

Theorem 4.7. *Let x be a k-dimensional vector, and let $\psi_i(\lambda, x)$, for each i, be a continuous function which satisfies the following conditions:*

a) $\psi_i(0,0) = 0$ for $i = 1, \ldots, k$;

b) $\psi_i(\lambda, x)$ *is differentiable at the point* $\lambda = 0$, $x = 0$, *i.e.,*

$$\left| \psi_i(\lambda, x) - \psi_i(0,0) - \lambda \frac{\partial \psi_i(0,0)}{\partial \lambda} - \sum_{j=1}^{k} \frac{\partial \psi_i(0,0)}{\partial x_j} x_j \right| \le \bar{r}\left(\sqrt{\lambda^2 + \|x\|^2}\right),$$

where $\|x\| = \left(\sum_{j=1}^{k} x_j^2\right)^{1/2}$, *and* $\bar{r}(z) z^{-1} \to 0$ *as* $z \to 0$;

c) $\dfrac{\partial \psi_i(0,0)}{\partial \lambda} = 0$ for $i = 1, \ldots, k$;

d) *the matrix* $\partial_x \psi = \left\{ \dfrac{\partial \psi_i}{\partial x_j} \right\}$, $i = 1, \ldots, k$, $j = 1, \ldots, k$, *is non-singular.*

Then the system of equations

$$\psi_i(\lambda, x) = 0, \quad i = 1, \ldots, k,$$

has a solution for sufficiently small λ, and there exists a solution $x(\lambda)$ with the property that

$$\lim_{\lambda \to 0} \frac{\| x(\lambda) \|}{\lambda} = 0 .$$

(If we suppose that the functions $\psi_i(\lambda, x)$ are continuously differentiable in some neighborhood of the point $\lambda = 0$, $x=0$, then Theorem 4.7 coincides with the usual implicit function theorem [45]).

Proof. It is convenient to carry out the reasoning using vector notation. Thus we denote

$$\psi(\lambda, x) = (\psi_1(\lambda, x), \ldots, \psi_k(\lambda, x)) .$$

The hypotheses of the theorem imply that

$$\begin{aligned} \psi(\lambda, x) &= \partial_x \psi x + r(\lambda, x), \\ \| r(\lambda, x) \| &\leq \bar{r}\left(\sqrt{\lambda^2 + \|x\|^2}\right). \end{aligned} \tag{4.32}$$

Consider the mapping

$$g(\lambda, x) = x - (\partial_x \psi)^{-1} \psi(\lambda, x) .$$

Using (4.32), we obtain

$$g(\lambda, x) = -(\partial_x \psi)^{-1} r(\lambda, x) ,$$

$$\| g(\lambda, x) \| \leq \bar{r}\left(\sqrt{\lambda^2 + \|x\|^2}\right) \| (\partial_x \psi)^{-1} \| . \tag{4.33}$$

We now note that, without loss of generality, we can consider $\bar{r}(z)$ to be a non-decreasing function of z. Indeed, if this is not the case, $\bar{r}(z)$ can be replaced by

$$\omega(z) = \sup_{0 \le t \le z} \bar{r}(t) ,$$

so that $\omega(z) \ge \bar{r}(z)$. Here, obviously, $\omega(z)$ is a non-decreasing function, and

$$\frac{\omega(z)}{z} \to 0 .$$

Indeed, since $\bar{r}(t)\, t^{-1} \to 0$, for every $\varepsilon > 0$ there exists a $\delta > 0$ such that

$$\frac{\bar{r}(t)}{t} \le \varepsilon$$

whenever $t \le \delta$. Now if $z < \delta$, then

$$\frac{\omega(z)}{z} = \frac{1}{z} \sup_{0 \le t \le z} \bar{r}(t) \le \sup_{0 \le t \le z} \frac{r(t)}{t} \le \varepsilon .$$

Now set

$$\tau(\lambda) = \inf\left\{\tau :\ K\bar{r}\left(\sqrt{\lambda^2 + \tau^2}\right) \le \tau\right\},$$

where $K = \| (\partial_x \psi)^{-1} \|$. Since

$$K\bar{r}\left(\sqrt{\lambda^2 + \lambda^2}\right) = K\bar{r}\left(\lambda\sqrt{2}\right) \le \lambda ,$$

for sufficiently small λ, $\tau(\lambda) \le \lambda$ for all such λ. Moreover, by definition of an infimum, for every such λ there exists a

$\tau^*(\lambda)$ such that

$$\tau(\lambda) \leq \tau^*(\lambda) \leq \tau(\lambda) + \lambda^2 ,$$

and

$$K\bar{r}\left(\sqrt{\lambda^2 + (\tau^*(\lambda))^2}\right) \leq \tau^*(\lambda) . \tag{4.34}$$

Let us show that

$$\frac{\tau(\lambda)}{\lambda} \to 0 .$$

Indeed, by definition,

$$\tau(\lambda) - \lambda^2 < K\bar{r}\left(\sqrt{\lambda^2 + (\tau(\lambda) - \lambda^2)^2}\right) \leq K\bar{r}(\lambda + |\tau(\lambda) - \lambda^2|), \tag{4.35}$$

where we have made use of the fact that $\bar{r}(z)$ is a non-decreasing function and that

$$\sqrt{\lambda^2 + \omega^2} \leq \lambda + \omega$$

for positive λ and ω. Further, since $\tau(\lambda) \leq \lambda$ for small λ, the right-hand side of (4.35) can be estimated as follows:

$$K\bar{r}(\lambda + |\tau(\lambda) - \lambda^2|) \leq K\bar{r}(\lambda + \lambda) = K\bar{r}(2\lambda).$$

Thus, we obtain

$$\tau(\lambda) - \lambda^2 \leq K\bar{r}(2\lambda) ,$$

or

$$\frac{\tau(\lambda)}{\lambda} \leq 2K\frac{\bar{r}(2\lambda)}{2\lambda} + \lambda ,$$

i.e.,

$$\frac{\tau(\lambda)}{\lambda} \to 0$$

as $\lambda \to 0$. But then also

$$\frac{\tau^*(\lambda)}{\lambda} \to 0.$$

Thus, Inequality (4.34) holds for sufficiently small λ.

Now if

$$\|x\| \leq \tau^*(\lambda),$$

then, according to (4.33),

$$\|g(\lambda,x)\| \leq K\bar{r}\left(\sqrt{\lambda^2 + \|x\|^2}\right) \leq K\bar{r}\left(\sqrt{\lambda^2 + (\tau^*(\lambda))^2}\right) \leq \tau^*(\lambda).$$

This implies that the continuous mapping $g(\lambda,x)$ maps the ball

$$\|x\| \leq \tau^*(\lambda)$$

into itself. By virtue of the Brouwer fixed point theorem [1], we can now assert that $g(\lambda,x)$ has a fixed point, i.e., there exists a point $x(\lambda)$ such that

$$x(\lambda) = g(\lambda, x(\lambda)),$$

$$\|x(\lambda)\| \leq \tau^*(\lambda).$$

But the definition of $g(\lambda,x)$ implies that

$$\psi(\lambda, x(\lambda)) = 0,$$

i.e., the set of nonlinear equations under consideration has a solution. Moreover,

$$\frac{\|x(\lambda)\|}{\lambda} \leq \frac{\tau^*(\lambda)}{\lambda},$$

which implies that

$$\frac{\|x(\lambda)\|}{\lambda} \to 0$$

as $\lambda \to 0$. |||

CHAPTER V

NECESSARY CONDITIONS FOR AN EXTREMUM IN CONCRETE PROBLEMS

In this chapter, we shall illustrate how the general theory which has been developed in the preceding chapters may be applied to various extremal problems. The power of any general theory lies in the fact that it allows one to consider in a uniform way various particular problems and to obtain complete results by a single method.

A number of the problems which we shall consider in the sequel has been considered previously. A number of works has been devoted to these problems. As a rule, to investigate each of these problems, some specific method has been used, with each method applicable just for solving the given, narrow problem. We shall show that we can apply the general theory which has been developed to every problem which we shall consider in the sequel, and that we can at once obtain, without any additional, lengthy reasoning, the complete results. These results are often even more general than those obtained by special methods.

1. <u>The classical mathematical programming problem.</u> Let there be given continuously differentiable functions $f_i(x)$, $i = -m, \ldots, k$, on an n-dimensional space. We are looking for conditions which a point x_0 that is a solution to the following problem:

$$\min f_0(x),$$
$$f_i(x) \leq 0 \text{ for } i = -m, \ldots, -1,$$
$$f_i(x) = 0 \text{ for } i = 1, \ldots, k.$$

must satisfy.

Since the functions $f_i(x)$ are differentiable, they belong to the class of quasi-differentiable functions. Moreover, for these functions, the set $M_i(x_0)$ consists of a single vector which coincides with the gradient of $f_i(x)$ at x_0. We denote this gradient by $\partial_x f_i(x_0)$. Indeed, by a known theorem of analysis [44]:

$$\frac{\partial f_i}{\partial e} = (\partial_x f_i(x_0), e) .$$

It is easy to see that the conditions under which we can apply Theorem 4.2 are satisfied, where M is the entire space. Thus, K_M is also the entire space, and $K_N^* = \{0\}$.

Now Theorem 4.2 implies that there exist numbers

λ_i, with $\lambda_i \geq 0$ for $i \leq 0$, such that

$$\sum_{i=-m}^{k} \lambda_i \partial_x f_i(x_0) = 0,$$

$$\lambda_i f_i(x_0) = 0 \quad \text{for} \quad i \neq 0.$$

We have obtained the well-known Lagrange multiplier rule.

2. Mathematical programming with a continuum of constraints. We are to find the minimum of a continuously differentiable function $\mu(x)$ of an n-dimensional argument x, subject to the constraints

$$\mu(\alpha, x) \leq 0 \quad \text{for} \quad \alpha \in \Omega,$$

where $\mu(\alpha, x)$ is a function which is continuous with respect to α and x and which has a continuous gradient $\partial_x \mu(\alpha, x)$. Here, Ω is a compact set.

Let us introduce the function

$$\mu_{-1}(x) = \max_{\alpha \in \Omega} \mu(\alpha, x).$$

Then the just formulated problem is equivalent to the problem

$$\min_{\mu_{-1}(x) \leq 0} \mu(x).$$

According to Theorem 3.3, $\mu_{-1}(x)$ is a quasi-differentiable function, and, according to Theorem 3.4, $M_{-1}(x_0)$ for this function is given by the formula

$$M_{-1}(x_0) = \overline{\text{co}}\Big(\bigcup_{\alpha \in \Omega(x_0)} M(x_0, \alpha) \Big),$$

where $\Omega(x_0) = \{ \alpha : \alpha \in \Omega, \ \mu(\alpha, x_0) = \mu_{-1}(x_0) \}$. Since $\mu(\alpha, x)$ is continuously differentiable with respect to x, $M(x_0, \alpha) = \{ \partial_x \mu(\alpha, x_0) \}$, and we can write

$$M_{-1}(x_0) = \text{co}\Big(\bigcup_{\alpha \in \Omega(x_0)} \partial_x \mu(\alpha, x_0) \Big).$$

Here, the closure bar has been omitted because the convex hull of a compact set in a finite-dimensional space is compact, and therefore closed, and the set

$$\bigcup_{\alpha \in \Omega(x_0)} \partial_x \mu(\alpha, x_0)$$

is compact because $\Omega(x_0)$ is compact and $\partial_x \mu(\alpha, x_0)$ depends continuously on α.

On the basis of Theorem 4.2, we can assert that there exist non-negative constants λ_0 and λ_{-1}, not both zero, such that

$$\lambda_0 \partial_x \mu(x_0) + \lambda_{-1} c = 0 \text{ and } \lambda_{-1} \mu_{-1}(x_0) = 0,$$

where $c \in M_{-1}(x_0)$. But $M_{-1}(x_0)$ is the convex hull of the n-dimensional vectors $\partial_x \mu(\alpha, x_0)$, where $\alpha \in \Omega(x_0)$. This implies that the vector c can be represented in the form

$$c = \sum_{i=1}^{n+1} \lambda_i \partial_x \mu(\alpha_i, x_0),$$

where, for each i, $\alpha_i \in \Omega(x_0)$, and $\lambda_i \geq 0$, and $\sum_{i=1}^{n+1} \lambda_i = 1$. Thus, we finally obtain

$$\lambda_0 \partial_x \mu(x_0) + \sum_{i=1}^{n+1} \gamma_i \partial_x \mu(\alpha_i, x_0) = 0, \tag{5.1}$$

where $\gamma_i = \lambda_1 \lambda_i \geq 0$ and $\sum_{i=1}^{n+1} \gamma_i = \lambda_{-1}$. Let us formulate the result which we have just obtained.

Theorem 5.1. *In order that x_0 be a solution to the mathematical programming problem with a continuum of constraints, it is necessary that there exist constants $\lambda_0 \geq 0$ and $\gamma_i \geq 0$, not all zero, and points $\alpha_i \in \Omega(x_0)$, $i=1,\ldots,n+1$, such that Equation (5.1) holds and such that $\gamma_i = 0$ for every i if $\mu_{-1}(x_0) < 0$.*

Now suppose that $\mu(x)$ and $\mu(\alpha, x)$, for each fixed α, are functions which are convex in x, and that there exists an x_1 such that $\mu_{-1}(x_1) < 0$. Then the problem under consideration becomes a convex programming problem.

Applying Theorem 2.4 to the case of a single

constraint, we obtain that there exist a non-negative number λ_{-1} and a vector $c \in M_{-1}(x_0)$ such that

$$\partial_x \mu(x_0) = -\lambda_{-1} c ,$$

and this condition is both necessary and sufficient.

Now, arguing in the same way as before, we obtain the following corollary.

Corollary 1. If $\mu(x)$ and $\mu(\alpha, x)$, for each fixed α, are convex functionals, and $\mu_{-1}(x_1) < 0$ for some x_1, then, in order that x_0 be a solution to the problem with a continuum of constraints, it is necessary and sufficient that there exist numbers $\gamma_i \geq 0$ and points $\alpha_i \in \Omega(x_0)$ such that Condition (5.1) holds with $\lambda_0 = 1$.

Corollary 2. Under the assumptions of Corollary 1, in order that x_0 be a solution to the problem with a continuum of constraints, it is necessary and sufficient that there exist points $\alpha_i \in \Omega(x_0)$ such that x_0 is a solution to the problem

$$\left.\begin{aligned} &\min \mu(x), \\ &\mu(\alpha_i, x) \leq 0 \quad \text{for } i = 1, \ldots, n+1 . \end{aligned}\right\} \tag{5.2}$$

Indeed, relation (5.1) (with $\lambda_0 = 1$) is at the same time a necessary and sufficient condition (by virtue of Theorem

2.4) for x_0 to be a solution to Problem (5.2). |||

Thus, Corollary 2 shows that, in the convex case, the problem with a continuum of constraints can be reduced to another, specially chosen problem which has at most (n+1) constraints.

3. Theorems for minimax problems. Let there be given a convex, compact set M in a space B and a convex, bounded, weak* closed set M^* in the space B^*. Let us introduce the function $\varphi(x) = \max_{x^* \in M^*} x^*(x)$.

Let $x_0 \in M$ be a minimum point of $\varphi(x)$ on M. Then, according to Theorem 2.1, there exists an $x_0^* \in M(x_0)$, where $M(x_0)$ is the set of support functionals to $\varphi(x)$ at x_0, such that $x^* \in \Gamma_{x_0}^*$, where Γ_{x_0} is the cone defined by the relation

$$\Gamma_{x_0} = \{e: x_0 + \lambda e \in M \text{ for sufficiently small } \lambda > 0\}.$$

In particular, $e = x - x_0$ with $x \in M$, belongs to this cone.

Further, by Theorem 1.6,

$$M(x_0) = \{x^*: x^* \in M^*, x^*(x_0) = \varphi(x_0)\}.$$

Thus, there exists a functional $x_0^* \in M(x_0)$ such that

$$x_0^*(x - x_0) \geq 0 \text{ for all } x \in M$$

and

$$x_0^*(x_0) \geq x^*(x_0) \text{ for all } x^* \in M^*.$$

Therefore, the two preceding inequalities imply that

$$x_0^*(x) \geq x_0^*(x_0) \geq x^*(x_0) \quad \text{for all } x \in M \text{ and } x^* \in M^*. \qquad (5.3)$$

Theorem 5.2. *If M is a convex, compact set in B, and M^* is a convex, bounded, weak* closed set in B^*, then*

$$\min_{x \in M} \max_{x^* \in M^*} x^*(x) = \max_{x^* \in M^*} \min_{x \in M} x^*(x). \qquad (5.4)$$

Proof. The theorem is an immediate consequence of (5.3), since (5.3) (see [5]) is equivalent to (5.4). |||

Corollary 1. *If X and Y are compact sets in E^n, $k(x,y)$ is a continuous function of the variables $x \in X$ and $y \in Y$, and X^* and Y^* denote, respectively, the sets of all regular, positive measures μ on X and ν on Y such that $\mu(X) = \nu(Y) = 1$, then*

$$\min_{\mu \in X^*} \max_{\nu \in Y^*} \int k(x,y)\mu(dx)\nu(dy)$$
$$= \max_{\nu \in Y^*} \min_{\mu \in X^*} \int k(x,y)\mu(dx)\nu(dy).$$

Proof. Let B be the space of all continuous functionals on Y, let M be the set of all functionals c(y) which can be represented in the form

$$c(y) = \int k(x,y)\mu(dx),$$

and let $M^* = Y^*$. Then the corollary follows immediately from Theorem 5.2. |||

<u>Corollary 2.</u> <u>If X and Y are compact, convex sets in E^n, $k(x,y)$ is a continuous function of the arguments $x \in X$ and $y \in Y$, and k is, in addition, convex in x and concave in y, then</u>

$$\min_{x\in X} \max_{y\in Y} k(x,y) = \max_{y\in Y} \min_{x\in X} k(x,y) .$$

<u>Proof.</u> On the basis of the preceding corollary to Theorem 5.2, we can assert that there exist measures μ_0 and ν_0, with $\mu_0(X) = \nu_0(Y) = 1$, such that

$$\begin{aligned} \int k(x,y)\mu(dx)\nu_0(dy) &\geq \int k(x,y)\mu_0(dx)\nu_0(dy) \\ &\geq \int k(x,y)\mu_0(dx)\nu(dy). \end{aligned} \tag{5.5}$$

Now set

$$x_\mu = \int x\mu(dx) \text{ and } y_\nu = \int y\nu(dy).$$

Then, since $k(x,y)$ is convex in x and concave in y,

$$\int k(x,y)\mu_0(dx)\nu(dy) \geq \int k(x_{\mu_0}, y)\nu(dy), \tag{5.6}$$

$$\int k(x,y)\mu(dx)\nu_0(dy) \leq \int k(x, y_{\nu_0})\mu(dx). \tag{5.7}$$

We shall show that

$$\int k(x,y)\mu_0(dx)\nu_0(dy) = \int k(x_{\mu_0}, y)\nu_0(dy). \qquad (5.8)$$

Indeed, on the basis of Inequality (5.6), the right-hand side of the last equation cannot be greater than the left-hand side. Suppose that it is strictly less, so that

$$\int k(x,y)\mu_0(dx)\nu_0(dy) > \int k(x_{\mu_0}, y)\nu_0(dy).$$

If we now choose the measure which is concentrated at the point x_{μ_0} for μ in the left-hand side of (5.5), we obtain a contradiction.

In exactly the same way it can be shown that

$$\int k(x_{\mu_0}, y)\nu_0(dy) = k(x_{\mu_0}, y_{\nu_0}). \qquad (5.9)$$

Indeed, since k(x, y) is concave in y,

$$\int k(x_{\mu_0}, y)\nu_0(dy) \le k(x_{\mu_0}, y_{\nu_0}).$$

Suppose that strict inequality holds. It follows from (5.5), (5.6), and (5.8) that

$$\int k(x_{\mu_0}, y)\nu_0(dy) \ge \int k(x_{\mu_0}, y)\nu(dy).$$

Choosing for ν the measure which is concentrated at y_{ν_0}, we obtain a contradiction to our assumption.

Now, on the basis of (5.5)-(5.9), we obtain that

$$\int k(x, y_{\nu_0})\mu(dx) \geq k(x_{\mu_0}, y_{\nu_0}) \geq \int k(x_{\mu_0}, y)\nu(dy).$$

If we choose for μ and ν measures which are concentrated at x and y, respectively, then we obtain

$$k(x, y_{\nu_0}) \geq k(x_{\mu_0}, y_{\nu_0}) \geq k(x_{\mu_0}, y)$$

for all $x \in X$ and $y \in Y$. |||

4. Chebyshev approximation problems. Let there be given a function $\mu(x, \alpha)$, where x ranges over some set Ω, and α ranges over a set $\mathfrak{A}$. We suppose that Ω is a set in E^n and that $\mathfrak{A}$ is a compact set in E^s. The function $\mu(x, \alpha)$ is continuous and has a continuous gradient $\partial_x \mu(x, \alpha)$ with respect to x. We shall consider the problem of minimizing the function

$$\mu(x) = \max_{\alpha \in \mathfrak{A}} \mu(x, \alpha)$$

on a set Ω which has, at every point x, a non-empty, convex cone K_x:

$$K_x = \{e: x + \lambda e \in \Omega \text{ for sufficiently small } \lambda > 0\}.$$

By virtue of what we proved in Chapter 3, $\mu(x)$ is quasi-differentiable, and M(x) is defined for μ by the formulas

$$M(x) = \text{co}\left(\bigcup_{\alpha \in \mathcal{A}(x)} \partial_x \mu(x, \alpha) \right),$$

$$\mathcal{A}(x) = \{ \alpha : \alpha \in \mathcal{A},\ \mu(x, \alpha) = \mu(x) \}.$$

On the basis of the corollary to Theorem 4.1 and Theorem 4.2, we can assert that, if $\mu(x)$ achieves its minimum on Ω at x_0, then the following condition holds: There exists a vector $c_0 \in K^*_{x_0}$ such that $c_0 \in M(x_0)$. But, since we are dealing with an n-dimensional space, any element in the convex hull of a set can be represented as a convex combination of (n+1) vectors of this set. Thus, there exist numbers λ_i, $i=1,\ldots,n+1$, and points $\alpha_i \in \mathcal{A}(x_0)$, such that

$$c_0 = \sum_{i=1}^{n+1} \lambda_i \partial_x \mu(x_0, \alpha_i),\quad \sum_{i=1}^{n+1} \lambda_i = 1,\ \lambda_i \geq 0. \tag{5.10}$$

<u>Theorem</u> 5.3. <u>In order that the previously defined function $\mu(x)$ achieve its minimum at $x_0 \in \Omega$, it is necessary that there exist constants $\lambda_i \geq 0$ and points $\alpha_i \in \mathcal{A}(x_0)$, $i=1,\ldots,n+1$, such that</u>

$$\sum_{i=1}^{n+1} \lambda_i \partial_x \mu(x_0, \alpha_i) \in K^*_{x_0},\quad \sum_{i=1}^{n+1} \lambda_i = 1. \tag{5.11}$$

<u>If $\mu(x, \alpha)$ is convex in x for all α and Ω is convex, then the stated conditions are sufficient.</u>

<u>Proof.</u> Formula (5.11) is an immediate consequence of the relations $c_0 \in K^*_{x_0}$ and (5.10). Moreover, if the additional convexity assumptions are satisfied, then the sufficiency follows from the convexity of $\mu(x)$, so that Theorem 2.1 can be applied to the problem. Theorem 2.1 in this case is entirely equivalent to Theorem 4.2 as far as the necessary conditions are concerned. But this theorem in addition implies that, in the convex case, the necessary conditions are at the same time also sufficient. |||

<u>Corollary. If Ω coincides with the entire space, then Condition (5.11) of Theorem 3.3 can be written in the form</u>

$$\sum_{i=1}^{n+1} \lambda_i \partial_x \mu(x_0, \alpha_i) = 0 .$$

Indeed, if Ω is the entire space, then K_{x_0} also coincides with E^n, and, thus, $K^*_{x_0}$ contains only the single element 0. |||

The just obtained results stated in Theorem 5.3 and its corollary can be considered as necessary conditions for a non-linear Chebyshev approximation problem. Indeed, we shall now show that the usual results for the classical Chebyshev approximation problem can be obtained from these results.

V. Necessary Conditions for an Extremum in Concrete Problems

Let there be given continuous functions $\varphi_k(\alpha)$, $k = 1, \ldots, n$, and a continuous function $f(\alpha)$ on a compact set E in an m-dimensional space. It is required to approximate this latter function by a generalized polynomial

$$F(\alpha) = \sum_{k=1}^{n} x_k \varphi_k(\alpha) ,$$

whose coefficients x_i range over a closed, convex domain Ω. This approximation should be best in the sense that the quantity

$$\mu(x) = \max_{\alpha \in E} \left| f(\alpha) - \sum_{k=1}^{n} x_k \varphi_k(\alpha) \right|$$

should be minimized.

Let us represent $\mu(x)$ in the following slightly different form which is more convenient for our investigation:

$$\mu(x) = \max_{\substack{\alpha \in E \\ 1 \le \xi \le +1}} \xi \left(f(\alpha) - \sum_{k=1}^{n} x_k \varphi_k(\alpha) \right). \qquad (5.12)$$

It is easy to see that $\mu(x)$ is convex in x. Further, μ is the maximum over a compact set of the family of functions

$$\mu(x, \alpha, \xi) = \xi \left(f(\alpha) - \sum_{i=1}^{n} x_i \varphi_i(\alpha) \right) .$$

These functions are obviously continuously differentiable with respect to x, and

$$\partial_x \mu(x, \alpha, \xi) = -(\xi\varphi_1(\alpha), \xi\varphi_2(\alpha), \ldots, \xi\varphi_n(\alpha)).$$

On the basis of Theorem 5.3, in order that $x^0 \in \Omega$ yield a minimum for $\mu(x)$, it is necessary and sufficient that there exist numbers $\lambda_i \geq 0$ and pairs (α_i, ξ_i) such that

$$-\sum_{i=1}^{n+1} \lambda_i \xi_i \begin{pmatrix} \varphi_1(\alpha_i) \\ \vdots \\ \varphi_n(\alpha_i) \end{pmatrix} \in K^*_{x_0}, \quad \sum_{i=1}^{n+1} \lambda_i = 1, \tag{5.13}$$

and all the pairs (α_i, ξ_i) are such that they yield a maximum for the function (5.12) for $x = x^0$. This implies that

$$\left.\begin{aligned} \mu(x_0) &= \left| f(\alpha_i) - \sum_{k=1}^{n} x_k^0 \varphi_k(\alpha_i) \right|, \\ \xi_i &= \operatorname{sign}\left(f(\alpha_i) - \sum_{k=1}^{n} x_k^0 \varphi_k(\alpha_i) \right). \end{aligned}\right\} \tag{5.14}$$

The following theorem summarizes this result.

Theorem 5.4. *In order that the coefficients x_k^0, $k = 1, \ldots, n$, be a solution to the previously stated Chebyshev approximation problem, it is necessary and sufficient that there exist numbers $\lambda_i \geq 0$ and points $\alpha_i \in E$ such that the formulas (5.14) and (5.13) hold, where $K^*_{x_0}$ is the dual cone to K_{x_0}.*

Corollary. If Ω is the entire space, then Condition (5.13) can be written in the form

$$\sum_{i=1}^{n+1} \lambda_i \xi_i \begin{pmatrix} \varphi_1(\alpha_i) \\ \cdot \\ \cdot \\ \cdot \\ \varphi_n(\alpha_i) \end{pmatrix} = 0 . \qquad (5.15)$$

We shall now apply the result which we have just obtained to the case where E is a subset of the real line, and $\varphi_k(\alpha) = \alpha^{k-1}$. Let α_i be points at which

$$\mu(x^0) = \max_{\alpha \in E} \left| f(\alpha) - \sum_{k=1}^{n} x_k^0 \alpha^{k-1} \right|$$

$$= \left| f(\alpha_i) - \sum_{k=1}^{n} x_k^0 \alpha_i^{k-1} \right| , \qquad (5.16)$$

$$\xi_i = \operatorname{sign}\left(f(\alpha_i) - \sum_{k=1}^{n} x_k^0 \alpha_i^{k-1} \right),$$

and suppose that $\alpha_1 \leq \alpha_2 \leq \cdots \leq \alpha_{n+1}$. The existence of such numbers is guaranteed by Theorem 5.4. Then, according to (5.15), there exist numbers $\lambda_i \geq 0$, with $\sum_{i=1}^{n+1} \lambda_i = 1$, such that

$$\sum_{n-1}^{n+1} \lambda_i \xi_i \begin{bmatrix} \alpha_i^0 \\ \alpha_i^1 \\ \cdot \\ \cdot \\ \cdot \\ \alpha_i^{n-1} \end{bmatrix} = 0 . \qquad (5.17)$$

We now note that if some of the α_i are equal, then the corresponding vectors $\xi_i(\alpha_i^0, \alpha_i^1, \ldots, \alpha_i^{n-1})$ are also equal. Collecting together the coefficients of such vectors in (5.17), we conclude that there exist numbers α_i, with $\alpha_i < \alpha_{i+1}$, and numbers $\lambda_i' \geq 0$ such that (5.16) holds and such that

$$\sum_{i=1}^{r} \lambda_i' \xi_i \begin{pmatrix} \alpha_i^0 \\ \cdot \\ \cdot \\ \cdot \\ \alpha_i^{n-1} \end{pmatrix} = 0, \quad \sum_{i=1}^{r} \lambda_i' = 1,$$

where $r \leq n+1$ and r is the number of distinct points α_i. But, if the α_i are distinct, then the vectors $\xi_i(\alpha_i^0, \alpha_i^1, \ldots, \alpha_i^{n-1})$ are linearly independent if $r \leq n$, and this implies that $r = n+1$, i.e., there exist exactly n+1 points α_i such that (5.16) holds, and, moreover, $\lambda_i' = \lambda_i > 0$ for each i.

Thus, we can assume that all the α_i in (5.17) are distinct.

Now, let us consider (5.17) to be a system of equations for the unknowns $\lambda_i \xi_i$. Transposing the terms corresponding to $i = n+1$ to the right-hand sides of the equations, we obtain the system of equations

$$\sum_{i=1}^{n} \lambda_i \xi_i \begin{pmatrix} \alpha_i^0 \\ \cdot \\ \cdot \\ \cdot \\ \alpha_i^{n-1} \end{pmatrix} = -\lambda_{n+1} \xi_{n+1} \begin{pmatrix} \alpha_{n+1}^0 \\ \cdot \\ \cdot \\ \cdot \\ \alpha_{n+1}^{n-1} \end{pmatrix}$$

for the unknowns $\lambda_i \xi_i$, $i = 1, \ldots, n$. According to Cramer's rule, the solution of this system can be written in the form

$$\lambda_i \xi_i = -\lambda_{n+1} \xi_{n+1} \frac{D(\alpha_1, \ldots, \alpha_{i-1}, \alpha_{n+1}, \alpha_{i+1}, \ldots, \alpha_n)}{D(\alpha_1, \ldots, \alpha_n)},$$

where

$$D(\alpha_1, \ldots, \alpha_n) = \begin{vmatrix} \alpha_1^0 & \alpha_2^0 & \cdots & \alpha_n^0 \\ \alpha_1^1 & \alpha_2^1 & \cdots & \alpha_n^1 \\ \cdot & \cdot & \cdot & \cdot \\ \alpha_1^{n-1} & \alpha_2^{n-1} & \cdots & \alpha_n^{n-1} \end{vmatrix}.$$

The determinant $D(\alpha_1, \ldots, \alpha_n)$ is a Vandermonde determinant which, as is well known, is positive if $\alpha_1 < \alpha_2 < \cdots < \alpha_n$. Further, since an interchange of columns in a determinant results in a change in sign,

$$D(\alpha_1, \ldots, \alpha_{i-1}, \alpha_{n+1}, \alpha_{i+1}, \ldots, \alpha_n)$$
$$= (-1)^{n-i} D(\alpha_1, \ldots, \alpha_{i-1}, \alpha_{i+1}, \ldots, \alpha_n, \alpha_{n+1}).$$

Therefore,

$$\lambda_i \xi_i = -\lambda_{n+1} \xi_{n+1} (-1)^{n-i} \frac{D(\alpha_1, \ldots, \alpha_{i-1}, \alpha_{i+1}, \ldots, \alpha_n, \alpha_{n+1})}{D(\alpha_1, \ldots, \alpha_n)} .$$

Since the λ_i (for $i = 1, \ldots, n+1$) and both determinants are positive, it follows from the last formula that

$$\xi_i = -(-1)^{n-i} \xi_{n+1},$$

i.e., at the points $\alpha_1, \alpha_2, \ldots, \alpha_{n+1}$ at which $|f(\alpha) - \sum_{k=1}^{n} x_k^0 \alpha^{k-1}|$ achieves its maximum, the errors alternate in sign.

Thus, we have shown that a polynomial

$$F(\alpha) = \sum_{k=1}^{n} x_k^0 \alpha^{k-1}$$

will be a best approximation of a continuous function $f(\alpha)$ on a compact set E on the real line if and only if the deviation

$$\left| f(\alpha) - \sum_{k=1}^{n} x_k^0 \alpha^{k-1} \right|$$

achieves its maximum at at most n+1 points $\alpha_1 < \alpha_2 < \cdots < \alpha_{n+1}$, $\alpha_i \in E$, and, moreover, the signs of the errors at the points α_i alternate. This is a classical result in the theory of Chebyshev approximations.

Let us now return to the case where E is a compact set in the s-dimensional space E^s.

Theorem 5.4 and its corollary assert that the generalized polynomial

$$F^0(\alpha) = \sum_{k=1}^{n} x_k^0 \varphi_k(\alpha)$$

is a best approximation of the continuous function $f(\alpha)$ if and only if Conditions (5.14) and (5.15) are satisfied, i.e., if and only if the deviation $|f(\alpha) - F^0(\alpha)|$ achieves its maximum at points α_i, $i = 1, \ldots, r$, with $r \leq n+1$, and (5.15) holds. But if we replace E by the finite set $\tilde{E}$ consisting of the points α_i, $i = 1, \ldots, r$, then Conditions (5.14) and (5.15) are also necessary and sufficient conditions for $F^0(\alpha)$ to be a best approximation of $f(\alpha)$ on $\tilde{E}$. Thus, we have proved the following theorem.

<u>The fundamental theorem in the theory of Chebyshev approximations. A polynomial which is a best approximation of a continuous function $f(\alpha)$ on a compact set E is, at the same time, a best polynomial approximation of $f(\alpha)$ on $\tilde{E} = \{\alpha_1, \ldots, \alpha_r\}$, where $r \leq n+1$. In other words, there exist points $\alpha_i \in E$, $i = 1, \ldots, r$, with $r \leq n+1$, such that one of the polynomials which is a best approximation of $f(\alpha)$ on $\tilde{E}$ is, at the same time, a best polynomial approximation on the whole of E.</u>

V. Necessary Conditions for an Extremum in Concrete Problems

We shall now consider the general problem of the theory of Chebyshev approximation in a Banach space B. Let Ω be an arbitrary convex, closed set in B, and let $\mu(x) = \|x\|$. We shall find necessary and sufficient conditions for a point x_0 to yield a minimum for $\mu(x)$ on Ω.

Since

$$\mu(x) = \|x\| = \max_{x^* \in S^*} x^*(x) ,$$

$$S^* = \{x^*: x^* \in B^*, \quad \|x^*\| \leq 1\} ,$$

$\mu(x)$ is a convex functional, and the set of support functionals to μ at x consists of the functionals $x^* \in S^*$ with $x^*(x) = \|x\|$ (see Theorem 1.6).

On the basis of Theorem 2.1, we can assert that a point $x_0 \in \Omega$ yields a minimum for $\|x\|$ on Ω if and only if there exists a support functional to $\|x\|$ at x_0 such that

$$x_0^*(x_0) \leq x_0^*(x) \quad \text{for all} \quad x \in \Omega . \tag{5.18}$$

Taking into account the characterization of support functionals to $\|x\|$ which we have just stated, we can say that

$$x_0^*(x_0) = \|x_0\| \text{ and } \|x_0^*\| = 1. \tag{5.19}$$

Now suppose that Ω consists of all elements of the form

$$x = \bar{x} + \sum_{i=1}^{n} c_i x_i ,$$

where the vector $c = (c_1, \ldots, c_n)$ ranges over some convex domain C. Then, minimizing

$$\| x \| = \left\| \bar{x} + \sum_{i=1}^{n} c_i x_i \right\|$$

with respect to $c \in C$ is equivalent to finding a best approximation of $\bar{x}$ in the form of a linear combination of the elements x_i with the coefficients lying in a given domain.

Theorem 5. 5. *A vector $c^0 \in C$ yields a minimum for*

$$\left\| \bar{x} + \sum_{i=1}^{n} c_i x_i \right\|$$

if and only if there exists a functional x_0^* such that

$$x_0^*\left(\bar{x} + \sum_{i=1}^{n} c_i^0 x_i \right) = \left\| \bar{x} + \sum_{i=1}^{n} c_i^0 x_i \right\|, \quad \| x_0^* \| = 1 \qquad (5.20)$$

and

$$\sum_{i=1}^{n} x_0^*(x_i)(c_i^0 - c_i) \leq 0 \quad \text{for all } c \in C. \qquad (5.21)$$

Proof. In the case under consideration, Ω consists of all elements of the form $\bar{x} + \sum c_i x_i$, where $c \in C$. The assertion of the theorem is now an immediate consequence of

(5.18) and (5.19), once we substitute into these relations the following formulas for x and x_0:

$$x = \bar{x} + \sum_{i=1}^{n} c_i x_i, \quad x_0 = \bar{x} + \sum_{i=1}^{n} c_i x_i^0 .$$

<u>Corollary. If</u> $C = E^n$, <u>then Condition</u> (5.21) <u>may be replaced by the condition</u>

$$x_0^*(x_i) = 0, \qquad i = 1, \ldots, n . \tag{5.22}$$

Indeed, if $C = E^n$, then the numbers c_i in the left-hand side of (5.21) may be chosen arbitrarily and Inequality (5.21) will then hold only if Equations (5.22) hold. |||

5. <u>A linear optimal control problem with phase constraints.</u> We shall consider an object whose behavior is described by the system of differential equations

$$\frac{dx}{dt} = Ax + Bu, \quad 0 \leq t \leq T , \tag{5.23}$$

where x is an n-dimensional vector, A is an $n \times n$-matrix, and B is an $n \times r$-matrix.

The control u(t) is a measurable function, and, for every t, u(t) belongs to a compact, convex set $U \subset E^r$. We shall call such controls <u>admissible.</u>

We wish to find an admissible control u(t) such that the

corresponding trajectory satisfies the conditions

$$x(0) - x^0 = 0,$$

$$x(T) - x^1 = 0,$$

$$\mu_{-1}(x(\cdot)) = \max_{0 \leq t \leq T} g_{-1}(x(t)) \leq 0,$$

and such that the functional

$$\mu_0(x(\cdot)) = \max_{0 \leq t \leq T} g_0(x(t))$$

achieves its minimum. Here, $g_0(x)$ and $g_{-1}(x)$ are continuously differentiable functions of x.

Let us study the behavior of the functionals μ_0 and μ_{-1}. They are defined on the space C^n of all n-dimensional vector valued continuous functions. Since every trajectory of System (5.23) depends on the choice of the admissible control, μ_0 and μ_{-1} are functionals on the space of controls (through this dependence).

It is not difficult to see, by virtue of certain properties of Stieltjes integrals, that

$$\mu_i(x(\cdot)) = \max_{\sigma \in M^*} \int_0^T g_i(x(t))\, d\sigma(t),$$

where M^* is the set of all non-decreasing functions which satisfy the conditions $\sigma(0) = 0$ and $\sigma(T) = 1$.

With every function $x(t)$ we can associate the functions $g_{-1}(x(t))$ and $g_0(x(t))$ which are also continuous functions in the space C^1. In this way, we can define non-linear operators $\mathcal{G}_i: C^n \to C^1$ $(i = 0, -1)$.

If we now define the functional ν on C^1 by

$$\nu(y(\cdot)) = \max_{\sigma \in M^*} \int_0^T y(t)\, d\sigma(t),$$

then we can write

$$\mu_i(x(\cdot)) = \nu \mathcal{G}_i(x(\cdot)).$$

Now, noting that [2] M^* is a convex, weak* closed, bounded set in $(C^1)^*$, and applying Theorems 1.6 and 3.1, we conclude that $\mu_i(x(\cdot))$ is a quasi-differentiable functional. Moreover,

$$\frac{\partial \mu_i(x^0(\cdot))}{\partial e} = \max_{\sigma \in M^*(x_0(\cdot))} \int_0^T (\partial_x g_i(x^0(t)), e(t))\, d\sigma ,$$

where $M^*(x^0(\cdot))$ is the set of all functions $\sigma \in M^*$ such that

$$\int_0^T g_i(x^0(t))\, d\sigma(t) = \mu_i(x^0(\cdot)),$$

$e(t)$ is an arbitrary element in C^n, and $\partial_x g_i(x) = (\partial g_i/\partial x_1, \partial g_i/\partial x_2, \ldots, \partial g_i/\partial x_n)$.

We shall now show that $\mu_i(x(\cdot))$ is a quasi-differentiable functional, not only on the space of continuous functions, but also on the space of controls. In order to do this, we appeal to the well-known Cauchy formula [46] to express the solution of System (5.23) in terms of the control. Namely, by this formula,

$$x(t) = \Phi(t)x_0 + \Phi(t)\int_0^t \Phi^{-1}(\tau)Bu(\tau)d\tau ,$$

where $\Phi(t)$ is the matrix function which satisfies the system of equations

$$\frac{d\Phi}{dt} = A\Phi , \quad \Phi(0) = I,$$

and I is the identity matrix.

By virtue of this formula, if a control is given an increment of $E(t)$, then the corresponding trajectory will receive an increment of

$$e(t) = \Phi(t)\int_0^t \Phi^{-1}(\tau)BE(\tau)d\tau . \tag{5.24}$$

If we now consider $\mu_i(x(\cdot))$ to be a composite functional on the control space, then this functional turns out to be quasi-differentiable, and

$$\frac{\partial\mu(x^0(\cdot))}{\partial E} = \frac{\partial\mu(x^0(\cdot))}{\partial e}$$

$$= \max_{\sigma \in M^*(x^0(\cdot))} \int_0^T \left(\partial g_i(x^0(t)), \Phi(t)\int_0^t \Phi(\tau)BE(\tau)d\tau \right) d\sigma(\tau).$$

Further, by virtue of the Cauchy formula, the constraint $x(T) = x^1$ defines the constraint

$$\Phi(T)x_0 + \Phi(T)\int_0^T \Phi^{-1}(t)Bu(t)dt - x^1 = 0,$$

which is equivalent to imposing constraints in terms of n linear functionals.

Now let us apply Theorem 4.2 to the problem under consideration. To do this, we choose for M the set of all admissible controls. Then, by virtue of the convexity of U and the linearity of the equality-type constraints, we can choose

$$K_M = \{E(t): E(t) = \lambda(u(t) - u^0(t)),$$
$$u(t) \text{ is an admissible control}, \lambda \geq 0\},$$

where $u^0(t)$ is an optimal control in the problem under consideration.

Further, it is easy to see, by the definition of a dual cone, that the conclusion of Theorem 4.2 can be restated in the following equivalent form: There exist numbers λ_i, not all zero, and functionals $x_i^* \in M(x_0)$ such that

$$\sum_{i=-m}^{k} \lambda_i x_i^*(e) \geq 0 \quad \text{for all} \quad e \in K_M,$$

$$\lambda_i \geq 0 \quad \text{for} \quad i \leq 0, \quad \lambda_i \varphi_i(x_0) = 0 \text{ for } i \neq 0.$$

By the definition of a directional differential, the sets $M_i(x_0)$ $(i = 0 \text{ or } -1)$ in the present problem consist of all functionals x_i^* of the form

$$\int_0^T \left(\partial_x g_i(x^0(t)),\ \Phi(t) \int_0^t \Phi^{-1}(\tau) BE(\tau) d\tau \right) d\sigma(t),$$

where $\sigma(t) \in M^*(x^0(\cdot))$.

Since the equality-type constraints are linear, the functional x^* has the form

$$\Phi(T) \int_0^T \Phi^{-1}(t)\, BE(t)\, dt .$$

Applying Theorem 4.2, we can assert the following: In order that the control $u^0(t)$ and the corresponding trajectory $x^0(t)$ be optimal in the stated problem, it is necessary that there exist non-negative numbers λ_0 and λ_{-1}, a vector $\lambda = (\lambda_1, \ldots, \lambda_n)$ (with $\lambda_{-1}, \lambda_0, \lambda_1, \ldots, \lambda_n$ not all zero), and nondecreasing functions $\sigma_0(t)$ and $\sigma_{-1}(t)$ such that

$$\sum_{i=-1}^{0} \lambda_i \int_0^T \left(\partial g_i(x^0(t)),\ \Phi(t)\int_0^t \Phi^{-1}(\tau)BE(\tau)d\tau \right) d\sigma_i(t)$$

$$+ \left(\lambda, \Phi(T)\int_0^T \Phi^{-1}(t)BE(t)dt \right) \geq 0 \qquad (5.25)$$

for all $E(t) \in K_M$, or, equivalently, for all $E(t) = u(t) - u^0(t)$, where $u(t)$ is an admissible control. Moreover, the following conditions must hold:

$$\left.\begin{aligned} &\sigma_i(0) = 0, \quad \sigma_i(T) = 1, \\ &\int_0^T g_i(x^0(t))d\sigma_i(t) = \mu_i(x^0(t)) \text{ for } i = -1 \text{ and } 0, \\ &\lambda_{-1}\mu_{-1}(x^0(\cdot)) = 0. \end{aligned}\right\} \qquad (5.26)$$

We shall not tire the reader by carrying out in detail the somewhat tedious manipulations which amount to interchanging the order of integration in (5.25). After such manipulations, Inequality (5.25) reduces to the form

$$\int_0^T (\psi(\tau), Bu(\tau))d\tau \geq \int_0^T (\psi(\tau), Bu^0(\tau))d\tau , \qquad (5.27)$$

where

$$\psi(\tau) = \Phi^{-1*}(\tau)\left[\sum_{i=-1}^{0} \lambda_i \int_\tau^T \Phi^*(t)\partial_x g_i(x^0(t)d\sigma_i(t) + \Phi^*(T)\lambda \right], \qquad (5.28)$$

and the asterisk after the matrix denotes transposition. But it follows from Condition (5.27) that the optimal control must satisfy the condition

$$(\psi(\tau), Bu^0(\tau)) = \min_{\nu \in U} (\psi(\tau), B\nu) \qquad (5.29)$$

for almost all τ with $0 \leq \tau \leq T$.

The following theorem summarizes the results which we have obtained.

Theorem 5.6. *In order that an admissible control $u^0(t)$ be a solution to the problem formulated at the beginning of this section, it is necessary that there exist non-negative numbers λ_0 and λ_{-1}, a vector λ, and non-decreasing functions $\sigma_i(t)$ such that Conditions (5.29) hold, where $\psi(t)$ is defined by (5.28), and the numbers λ_i and λ and the functions $\sigma_i(t)$ satisfy Conditions (5.26).*

Remark 1. It is easy to see that if g_0 and g_{-1} are convex functions, then the problem which we have just considered is a convex programming problem, and, if $\lambda_0 > 0$, then the conditions of Theorem 5.6 are also sufficient.

Remark 2. Let us note some properties of $\psi(\tau)$. At every point where the functions $\sigma_0(\tau)$ and $\sigma_{-1}(\tau)$ are differentiable, i.e., almost everywhere (since $\sigma_0(\tau)$ and $\sigma_{-1}(\tau)$ are non-decreasing functions), $\psi(\tau)$ has a derivative given by the formula

$$\frac{d\psi}{d\tau} = -A^*\psi - \sum_{i=-1}^{0} \lambda_i \partial_x g_i(x^0(\tau)) \frac{d\sigma_i}{d\tau}.$$

This derivative can be obtained simply by differentiating (5.28). At every point of discontinuity of $\sigma_0(\tau)$ and $\sigma_{-1}(\tau)$, $\psi(\tau)$ has a jump, and, by definition of a Stieltjes integral,

$$\psi(\tau -0) - \psi(\tau +0) = \sum_{i=-1}^{0} \lambda_i \Delta\sigma_i(\tau)\partial g_i(x^0(\tau)).$$

Here, $\Delta\sigma_i(\tau)$ is the magnitude of the jump of σ_i at τ.

In conclusion, we note that, in the preceding problem, we can impose various more complicated constraints on the right-hand endpoint of the trajectory and on the phase coordinates. Such constraints can be taken into account much as we did before, without any particular difficulties.

6. <u>A duality principle in convex programming.</u> Let L be a linear space, let $\varphi_i(x)$, $i=-m, -(m-1), \ldots, -1, 0, \ldots, k$ be functionals which are convex for each $i \leq 0$ and linear for each $i > 0$, and let X be a convex set. We wish to solve the problem

$$\left.\begin{aligned} &\min \varphi_0(x), \\ &\varphi_i(x) \le 0 \text{ for } i < 0, \\ &\varphi_i(x) = 0 \text{ for } i > 0, \\ &x \in X. \end{aligned}\right\} \tag{5.30}$$

Let us consider the following set in (m+k+1)-dimensional space:

$$M = \{ z: z \in E^{m+k+1}, \; z_i = \varphi_i(x) \text{ for } -m \le i \le k, \; x \in X \}.$$

In what follows, we shall assume that M is compact.

We set

$$\mu(u) = \max_{x \in X} \sum_{i=-m}^{k} u_i \varphi_i(x) = \max_{z \in M} \sum_{i=-m}^{k} u_i z_i. \tag{5.31}$$

We shall consider the problem of minimizing this function on the set

$$\Omega = \{ u: u_0 = -1, \; u_i \le 0 \text{ for } i < 0 \}.$$

The function $\mu(u)$ is convex. Since the function

$$\mu(u, z) = \sum_{i=-m}^{k} u_i z_i$$

is differentiable with respect to u,

$$\partial_u \mu(u, z) = z, \qquad z = (z_{-m}, \ldots, z_0, \ldots, z_k),$$

then, according to Theorem 3.4 and Remarks 1, 2, and 3 to Theorem 3.2, the set of support functionals to $\mu(u)$ at u^0 is the convex hull of the set of all vectors z which satisfy the condition

$$\sum_{i=-m}^{k} u_i^0 z_i = \mu(u^0), \quad z \in M,$$

or, equivalently, the convex hull of the set of all vectors $\varphi(x)$ such that

$$\sum_{i=-m}^{k} u_i^0 \varphi_i(x) = \mu(u^0), \quad x \in X,$$

$$\varphi(x) = (\varphi_{-m}(x), \ldots, \varphi_0(x), \ldots, \varphi_k(x)). \tag{5.32}$$

Thus, every vector in $M(u^0)$ can be represented in the form

$$\sum_{j=1}^{l} \lambda_j \varphi(x_j), \quad \sum_{j=1}^{l} \lambda_j = 1, \quad \lambda_j \geq 0,$$

where every x_j satisfies Condition (5.32).

Now consider the set Γ_{u^0} of all "admissible directions for Ω at u^0". It is easily seen that Γ_{u^0} consists of all vectors $e = (e_{-m}, \ldots, e_0, \ldots, e_k)$ such that:

e_i is arbitrary if $u_i^0 < 0$ and $i < 0$;

$e_i \leq 0$ if $u_i^0 = 0$ and $i < 0$;

$e_0 = 0$;

e_i is arbitrary if $i > 0$.

Therefore, the dual cone consists of all vectors e^*

such that

$$e_i^* = 0 \text{ if } u_i^0 < 0 \text{ and } i < 0;$$

$$e_i^* \leq 0 \text{ if } u_i^0 = 0 \text{ and } i < 0;$$

$$e_0^* \text{ is arbitrary;}$$

$$e_i^* = 0 \text{ if } i > 0.$$

According to Theorem 2.1, the following condition holds at any point $u^0 \in \Omega$ at which $\mu(u)$ achieves its minimum on Ω:

$$M(u^0) \cap \Gamma^*_{u^0} \neq \emptyset .$$

This means that there exist points $x_j \in X$, $j=1, \ldots, \ell$, and numbers $\lambda_j \geq 0$, with $\sum_{j=1}^{\ell} \lambda_j = 1$, such that the vector

$$c = \sum_{j=1}^{l} \lambda_j \varphi(x_j)$$

belongs to $\Gamma^*_{u^0}$, and such that each x_j satisfies Equation (5.32). But, since $c \in \Gamma^*_{u^0}$, this means, by virtue of what we have just said about $\Gamma^*_{u^0}$, that

$$\left.\begin{aligned} &\sum_{j=1}^{l} \lambda_j \varphi_i(x_j) = 0 \quad \text{if} \quad u_i^0 < 0 \quad \text{and} \quad i < 0, \\ &\sum_{j=1}^{l} \lambda_j \varphi_i(x_j) \leq 0 \quad \text{if} \quad u_i^0 = 0 \quad \text{and} \quad i < 0, \\ &\sum_{j=1}^{l} \lambda_j \varphi_i(x_j) = 0 \quad \text{if} \quad i > 0. \end{aligned}\right\} \qquad (5.33)$$

Now consider the point

$$x_0 = \sum_{j=1}^{l} \lambda_j x_j.$$

Since X is convex and $x_j \in X$, also $x_0 \in X$. Further, by the convexity of the functionals $\varphi_i(x)$ for $i \leq 0$ and their linearity for $i > 0$, we obtain, on the basis of (5.33), that

$$\left.\begin{aligned} \varphi_i(x_0) &\leq \sum_{j=1}^{l} \lambda_j \varphi_i(x_j) \leq 0 \text{ for } i < 0, \\ \varphi_i(x_0) &= \sum_{j=1}^{l} \lambda_j \varphi_i(x_j) = 0 \text{ for } i > 0. \end{aligned}\right\}$$

Finally,

$$\varphi_0(x_0) \leq \sum_{j=1}^{l} \lambda_j \varphi_0(x_j).$$

Thus, we have shown that x_0 satisfies all of the constraints of the original convex programming problem.

We shall now show that x_0 is a solution to this problem, i.e., that it yields a minimum for $\varphi_0(x)$.

Since $u_i^0 \leq 0$ for $i \leq 0$ and the functionals $\varphi_i(x)$ are convex for $i \leq 0$ and linear for $i > 0$,

$$\sum_{i=-m}^{k} u_i^0 \varphi_i(x_0) \geq \sum_{i=-m}^{k} u_i^0 \sum_{j=1}^{l} \lambda_j \varphi_i(x_j)$$
$$= \sum_{j=1}^{l} \lambda_j \sum_{i=-m}^{k} u_i^0 \varphi_i(x_j) = \sum_{j=1}^{l} \lambda_j \mu(u^0) = \mu(u^0),$$

where we have made use of the fact that x_j satisfies (5.32). But, since $x_0 \in X$, it follows from the definition of $\mu(u)$ that

$$\sum_{i=-m}^{k} u_i^0 \varphi_i(x_0) \leq \mu(u^0).$$

Thus,

$$\sum_{i=-m}^{k} u_i^0 \varphi_i(x_0) = \mu(u^0) = \sum_{i=-m}^{k} u_i^0 \varphi_i(x_j).$$

Multiplying each of the last relations by λ_j and summing, and taking into account (5.33) and the relation $\sum_{i=1}^{\ell} \lambda_j = 1$, we conclude that

$$\sum_{i=-m}^{k} u_i^0 \varphi_i(x_0) = \sum_{i=-m}^{k} u_i^0 \Big(\sum_{j=1}^{\ell} \lambda_j \varphi_i(x_j) \Big) = u_0^0 \sum_{j=1}^{\ell} \lambda_j \varphi_0(x_j).$$

Because $\varphi_0(x)$ is convex and $u_0^0 = -1$, we obtain, from the last equation, that

$$\sum_{i=-m}^{k} u_i^0 \varphi_i(x_0) \leq u_0^0 \varphi_0(x_0),$$

or that

$$\sum_{\substack{i=-m \\ i \neq 0}}^{k} u_i^0 \varphi_i(x_0) \leq 0. \tag{5.34}$$

But we have already shown that $\varphi_i(x_0) \leq 0$ for $i < 0$ and that $\varphi_i(x_0) = 0$ for $i > 0$.

Because $u_i^0 \le 0$ for $i < 0$, each term in the left-hand side of (5.34) turns out to be non-negative. Thus, Inequality (5.34) can hold only if

$$u_i^0 \varphi_i(x_0) = 0 \quad \text{for} \quad i \ne 0.$$

Now let x be an arbitrary point which satisfies the constraints (5.30). Then, by definition of $\mu(u^0)$,

$$-\varphi_0(x_0) = \sum_{i=-m}^{k} u_i^0 \varphi_i(x_0) = \mu(u^0)$$

$$\ge -\varphi_0(x) + \sum_{\substack{i=-m \\ i \ne 0}}^{k} u_i^0 \varphi_i(x) \ge -\varphi_0(x),$$

since $u_i^0 \le 0$ for $i < 0$, $\varphi_i(x) \le 0$ for $i < 0$, and $\varphi_i(x) = 0$ for $i > 0$. Thus, the inequality $\varphi_0(x_0) \le \varphi_0(x)$ holds. This inequality shows that x_0 is a solution to the original convex programming problem.

We have proved the following theorem.

Theorem 5.7. *Let the set M which we have defined be compact, and let u^0 be a minimum point of $\mu(u)$ (defined by (5.31)) on Ω. Then a solution to the original convex programming problem (5.30) can be found among the points which satisfy the condition*

$$\sum_{i=-m}^{k} u_i^0 \varphi_i(x) = \mu(u^0), \quad x \in X.$$

What is the basic essence of the result which we have just obtained? This result shows that we can associate another problem with the original convex programming problem. This second problem has a more complicated criterion function, but has simple constraints, and solving it is equivalent to solving the original problem. It is known from the classical theory of linear programming that the dual problem sometimes turns out to be more convenient to solve than the original one. Incidentally, if we apply the result which we have obtained to an ordinary linear programming problem, then we indeed obtain the problem which is dual in the sense of linear programming theory [32].

Finally, we point out the following important peculiarity of the dual problem: it is finite-dimensional, whether or not the original problem is. It is precisely this fact which is widely made use of in a number of algorithms for solving linear optimal control problems.

7. Systems of convex inequalities. Helly's theorem. We shall show how the results in the theory of necessary conditions for an extremum which we have obtained may be

applied to investigate conditions regarding the compatibility of systems of convex inequalities.

Let L be a linear space and let $\varphi_i(x)$, for $i = 1, \ldots, n$, be convex functionals on L. We shall consider the system of inequalities

$$\varphi_i(x) \leq 0, \quad i = 1, \ldots, n, \quad x \in X. \tag{5.35}$$

where X is a convex set in L. Let

$$\varphi(x) = (\varphi_1(x), \ldots, \varphi_n(x)),$$

$$M = \{ z: z \in E^n, \quad z = \varphi(x), \; x \in X \}.$$

Suppose that M is compact. Then there exists the following bounded function

$$\mu(\lambda) = \max_{z \in M} (\lambda, z) = \max_{x \in X} (\lambda, \varphi(x)).$$

Let us consider this function on the set

$$\Omega = \left\{ \lambda: \lambda \leq 0, \; \sum_{i=1}^{n} \lambda_i = -1 \right\}.$$

This is a convex function on Ω which achieves its minimum at some point λ^0.

Theorem 5.8. If the set

$$M = \{ z: z \in E^n, \; z = \varphi(x); \; x \in X \}$$

is compact, then System (5.35) has a solution if and only if the function

$$\mu(\lambda) = \max_{x\in X} (\lambda, \varphi(x))$$

is non-negative for all $\lambda\in\Omega$.

Proof. We have already dealt more than once with functions of the form $\mu(\lambda)$, particularly in the preceding section of this chapter. We showed there that the set $M(\lambda)$ of support functionals to this function is the convex hull of all vectors $\varphi(x)$ which satisfy the condition

$$(\lambda, \varphi(x)) = \mu(\lambda), \quad x\in X . \tag{5.36}$$

Now suppose that $\lambda^0\in\Omega$ is a minimum point of $\mu(\lambda)$ and that $\mu(\lambda^0)$ is non-negative. On the basis of the corollary to Theorem 2.1, since the space is finite-dimensional, there must exist at λ^0 a vector $c^0\in M(\lambda)$ such that

$$(\lambda^0, c^0) \leq (\lambda, c^0) \quad \text{for all} \quad \lambda\in\Omega ,$$

i.e., by the definition of Ω,

$$(\lambda^0, c^0) = \min_{\lambda\in\Omega} (\lambda, c^0) = - \max_{1\leq i\leq n} c_i^0 . \tag{5.37}$$

Further, according to what was said about the structure of $M(\lambda)$, there exist numbers γ_j and points x_j satisfying

(5.36) such that

$$c^0 = \sum_{j=1}^{k} \gamma_j \varphi(x_j), \quad \sum_{j=1}^{k} \gamma_j = 1, \quad \gamma_j \geq 0.$$

It follows from this that

$$(\lambda^0, c^0) = \sum_{j=1}^{k} \gamma_j (\lambda^0, \varphi(x_j)) = \mu(\lambda^0) \geq 0.$$

Taking into account (5.37) and the formula for c^0, this yields

$$\max_i \sum_{j=1}^{k} \gamma_i \varphi_i(x_j) \leq 0.$$

Now let

$$x_0 = \sum_{j=1}^{k} \gamma_j x_j .$$

Since the functionals $\varphi_i(x)$ are convex,

$$\varphi_i(x_0) \leq \sum_{j=1}^{n} \gamma_j \varphi_i(x_j) \leq \max_i \sum_{j=1}^{n} \gamma_i \varphi_i(x_j) \leq 0,$$

i.e., x_0 is a solution of System (5.35).

The sufficiency part of the theorem has been proved. The necessity part is obvious because if $\varphi_i(x_0) \leq 0$ for some $x_0 \in X$, then $\mu(\lambda) = \max_{x \in X} (\lambda, \varphi(x)) \geq (\lambda, \varphi(x_0)) \geq 0$ for $\lambda \in \Omega$. |||

We shall now consider the finite-dimensional case

where $L = E^p$.

Theorem 5.9. Let X be a compact set in E^p. Then for System (5.35) to be compatible, it is necessary and sufficient that every subsystem of (5.35) which consists of at most p+1 inequalities be compatible.

Proof. The necessity is again obvious. Let us show the sufficiency. To do this, we consider the problem

$$\min \mu(x), \quad x \in X ,$$

where

$$\mu(x) = \max_{1 \leq i \leq n} \varphi_i(x) .$$

On the basis of Theorem 1.4, the set of support functionals to $\mu(x)$ consists of all p-dimensional vectors x^* which can be represented in the form

$$x^* = \sum_{i \in I(x_0)} \lambda_i x_i^* ,$$

where

$$I(x_0) = \{i: 1 \leq i \leq n, \ \varphi_i(x_0) = \mu(x_0)\},$$

x_i^* is an arbitrary support functional to $\varphi_i(x)$ at x, and

$$\sum_{i \in I(x_0)} \lambda_i = 1, \quad \lambda_i \geq 0.$$

We note that, since $x \in E^p$ and $(E^p)^* = E^p$, we can talk about functionals $x^* \in (E^p)^*$ as if they were p-dimensional vectors.

Now, if $M_i(x_0)$ denotes the set of support functionals to $\varphi_i(x)$ at x_0, then the result which we have stated concerning the structure of the set of support functionals to $\mu(x)$ can also be formulated as follows:

$$M(x_0) = \mathrm{co}\left(\bigcup_{i \in I(x_0)} M_i(x_0) \right).$$

Incidentally, this formula also follows immediately from Theorem 3.4. In this connection, the bar over co has been deleted because the sets $M_i(x_0)$ are closed, and, thus, their union (as the union of a finite number of closed sets) is also closed, and the convex hull of a compact set in a finite-dimensional space is also closed. Since the space under consideration is p-dimensional, every vector in the convex hull of a set can be represented as a convex combination of at most p+1 vectors of the original set. Therefore, every support functional (vector) to $\mu(x)$ can be represented in the form

$$x^* = \sum_{j=1}^{r} \lambda_j x_{i_j}^*, \text{ with } r \leq p+1, \tag{5.39}$$

where $i_j \in I(x_0)$ and $x^*_{i_j} \in M_{i_j}(x_0)$. A convex function $\mu(x)$ achieves its minimum on a compact set X at some point x_0. According to the corollary to Theorem 2.1, there exists a functional $x^*_0 \in M(x_0)$ such that

$$x^*_0(x_0) \leq x^*_0(x) \text{ for all } x \in X. \tag{5.40}$$

For our problem, this means that there exist an integer $r \leq p+1$, numbers $\lambda_j \geq 0$, and functionals $x^*_{i_j} \in M_{i_j}(x_0)$ such that x^*_0 (in (5.40)) can be represented in the form (5.39). Here, $i_j \in I(x_0)$ and

$$\sum_{j=1}^{r} \lambda_j = 1.$$

But, according to Theorem 1.4, x^*_0 is a support functional at x_0 to the convex function

$$\bar{\mu}(x) = \max_{1 \leq j \leq r} \varphi_{i_j}(x).$$

According to the corollary to Theorem 2.1, the fulfillment of (5.40) is necessary and sufficient for $\bar{\mu}(x)$ to achieve its minimum at x_0. Moreover,

$$\mu(x_0) = \bar{\mu}(x_0),$$

since $i_j \in I(x_0)$, and, thus, at x_0,

$$\varphi_{i_j}(x_0) = \mu(x_o) \text{ for } j = 1, \ldots, r,$$

by definition of $I(x_0)$.

By the hypotheses of the theorem, the system of inequalities

$$\varphi_{i_j}(x) \le 0, \ j = 1,\dots,r, \quad r \le p+1, \quad x \in X,$$

is compatible. This means that, for some point $x_1 \in X$,

$$\varphi_{i_j}(x_1) \le 0 \ \text{ for } j = 1,\dots,p,$$

and, consequently,

$$\bar{\mu}(x_1) \le 0.$$

Therefore

$$\mu(x_0) = \bar{\mu}(x_0) = \min_{x \in X} \bar{\mu}(x) \le \bar{\mu}(x_1) \le 0.$$

But it follows from the definition of $\mu(x_0)$ that

$$\varphi_i(x_0) \le \mu(x_0) \le 0,$$

i.e., x_0 is a solution to the system. |||

Let us show that the well-known theorem of Helly (see, e.g., [5] and [37]) is a consequence of the preceding result. Let there be given a finite system of convex, compact sets X_i, $i = 1,\dots,m$, in a p-dimensional space. We set

$$W_i(\psi) = \max_{x \in X_i} (\psi, x),$$

$$F_i(x) = \max_{\|\psi\| \leq 1} [(\psi, x) - W_i(\psi)].$$

It is easy to show, based on the separation theorem, that $x \in X_i$ if and only if

$$F_i(x) \leq 0,$$

where $F_i(x)$ is a convex function. Indeed, if $x_0 \in X_i$, then, by definition of $W_i(\psi)$,

$$(\psi, x_0) - W_i(\psi) \leq 0 \quad \text{for all } \psi.$$

Conversely, let $F_i(x_0) \leq 0$, i.e., let the last inequality hold for all ψ with $\|\psi\| = 1$. If we assume that $x_0 \notin X_i$, then, by the separation theorem, there exist a vector ψ_0 with $\|\psi_0\| = 1$ and a number $\delta > 0$ such that

$$(\psi_0, x) < (\psi_0, x_0) - \delta \quad \text{for all } x \in X_i,$$

i.e.,

$$\delta \leq (\psi_0, x_0) - W_i(\psi_0).$$

But this contradicts our assumption.

Thus, with every system of compact convex sets, we can associate the system of inequalities

$$F_i(x) \leq 0, \quad i=1,\ldots,m, \quad x \in X,$$

where the functions F_i are convex, and where we can choose for X a ball whose radius is sufficiently large that X contains all of the X_i. Obviously, the intersection of the X_i is non-empty if and only if the system is compatible. But, by Theorem 5.9, such a system is compatible if and only if every one of its subsystems consisting of at most p+1 inequalities is compatible. If we translate this into the language of convex sets, we obtain Helly's theorem.

Helly's theorem. If $X_1, \ldots, X_m$ are convex compact sets such that any p+1 of them have a non-empty intersection, then also the intersection of all of these sets is non-empty.

8. _The moment problem._ One of the very important mathematical problems, which has wide applications, is the moment problem, which can be described in the following way. Given a Banach space B and a finite number of elements $x_1, \ldots, x_n$ in B, we wish to find a functional $x^* \in B^*$ of minimum norm which satisfies the system of inequalities

$$x^*(x_i) - \alpha_i \leq 0 \text{ for } i = 1, \ldots, n,$$

where $\alpha_1, \ldots, \alpha_n$ are given numbers.

In order to relate the just-formulated problem to a problem on the solvability of a system of convex inequalities, we shall investigate the compatibility of the following system

$$x^*(x_i) - \alpha_i \leq 0 \text{ for } i=1,\dots,n, \quad \|x^*\| \leq \rho. \qquad (5.41)$$

Obviously, finding the smallest number ρ for which System (5.41) is compatible is equivalent to solving the moment problem.

In order to be able to apply Theorem 5.8, we must make sure that the set

$$M = \{ z: z \in E^n, \quad z_i = x^*(x_i) - \alpha_i, \ i=1,\dots,n, \|x^*\| \leq \rho \}$$

is compact. But, based on the fact that a convex, bounded, weak* closed set in B^* is weak* compact, we can assert that the set

$$S^*_\rho = \{ x^*: \|x^*\| \leq \rho \}$$

is weak* compact.

Indeed, if $\|x^*_0\| > \rho$, then there exists an element x_0 with $\|x_0\| \leq 1$ such that

$$x^*_0(x_0) > \rho .$$

We now choose an $\varepsilon > 0$ such that $(x^*_0(x_0) - \varepsilon) > \rho$. Then the neighborhood

$$W_0 = \{ x^*: |x^*(x_0) - x^*_0(x_0)| < \varepsilon \}$$

of x^*_0 does not intersect S^*, because, for $x^* \in W_0$,

$$x^*(x_0) > x^*_0(x_0) - \varepsilon > \rho ,$$

and, thus,

$$\|x^*\| = \sup_{\|x\| \leq 1} x^*(x) \geq x^*(x_0) > \rho .$$

Therefore, every point of the complement of S_ρ^* has a neighborhood which does not intersect S_ρ^*. This means that the complement of S_ρ^* is open, and, therefore, S_ρ^* is weak* closed. Further, by definition of the weak* topology, the mapping

$$z_i = x^*(x_i) - \alpha_i, \quad i=1,\dots,n,$$

is continuous. It follows from this that M itself, as the continuous image of a compact set, is compact in the usual Euclidean topology. Thus, it is sequentially compact, since in this topology the notions of sequential compactness and compactness coincide. (In connection with the notions and definitions which we have made use of here, see the introduction and also [1 , Lemma I. 5. 6 on p. 17, and Theorem V.4.2 and Corollary V.4.3 on p. 424]).

Since M is compact, we can appeal to Theorem 5.8.

For System (5.41), the function $\mu(\lambda)$ which was used in Theorem 5.8 has the form

$$\mu(\lambda) = \max_{\|x^*\| \leq \rho} \sum_{i=1}^{n} \lambda_i (x^*(x_i) - \alpha_i) \tag{5.42}$$

$$= \max_{\|x^*\| \leq \rho} x^*\left(\sum_{i=1}^{n} \lambda_i x_i\right) - \sum_{i=1}^{n} \lambda_i \alpha_i = \rho \left\| \sum_{i=1}^{n} \lambda_i x_i \right\| - \sum_{i=1}^{n} \lambda_i \alpha_i .$$

Theorem 5.10. *The system of inequalities (5.41) is compatible if and only if*

$$\rho \geq \sup_{\substack{\lambda_i \leq 0 \\ i=1,\ldots,n}} \frac{\sum_{i=1}^{n} \lambda_i \alpha_i}{\left\| \sum_{i=1}^{n} \lambda_i x_i \right\|} . \tag{5.43}$$

The previously formulated moment problem has a solution if the right-hand side of (5.43) (which we shall denote by σ) is finite. In this case, the minimum norm of a functional which is a solution to the problem is max (0, σ).

Proof. On the basis of Theorem 5.8, System (5.41) is compatible if and only if the function (5.42) is non-negative for all λ_i, $i=1,\ldots,n$, which satisfy the conditions

$$\lambda_i \leq 0, \quad \sum_{i=1}^{n} \lambda_i = -1 .$$

Formula (5.43) is simply another notation of this fact, which immediately follows from (5.42). Here, the condition

$\sum_{i=1}^{n} \lambda_i = -1$ can be omitted since the right-hand side of (5.43) does not change if we multiply all of the λ_i by the same positive constant. The conclusion of the theorem regarding the solvability of the moment problem follows from the previously mentioned relation between this problem and the problem of the solvability of (5.41), if we take into account the fact that the norm of a non-zero functional is positive. |||

In the literature, one often comes across a moment problem in a slightly different form. Namely, one is to find a functional of minimum norm which satisfies the conditions

$$x^*(x_i) - \alpha_i = 0 \quad \text{for } i=1,\ldots,n. \tag{5.44}$$

Obviously, this formulation is equivalent to the preceding one if we rewrite each of the equations in the form of two inequalities

$$x^*(x_i) - \alpha_i \le 0,$$
$$x^*(-x_i) - (-\alpha_i) \le 0,$$
$$\text{for } i=1,\ldots,n.$$

Therefore, the entire preceding theory can be applied to a moment problem in this new formulation. If we appeal to Theorem 5.10, taking into account that we have doubled our system of inequalities, we conclude that System (5.44) has a solution x^*, with $\|x^*\| \leq \rho$, only if

$$\rho \geq \sigma ,$$

where

$$\sigma = \sup_{\substack{\lambda^+ \leq 0 \\ \lambda^- \leq 0}} \frac{\sum_{i=1}^{n} (\lambda_i^+ - \lambda_i^-) \alpha_i}{\left\| \sum_{i=1}^{n} (\lambda_i^+ - \lambda_i^-) x_i \right\|},$$

or (if we introduce the notation $\lambda = \lambda^+ - \lambda^-$ and take into account that when λ^+ and λ^- range over the set of all non-positive numbers, λ ranges over the entire real axis)

$$\sigma = \sup_{\lambda} \frac{\sum_{i=1}^{n} \lambda_i \alpha_i}{\left\| \sum_{i=1}^{n} \lambda_i x_i \right\|}. \tag{5.45}$$

Thus, we can state the following corollary to Theorem 5.10.

Corollary. The minimum norm of a functional which satisfies the constraints (5.44) is equal to σ as defined by Eq. (5.45).

Remark. Obviously, if not all of the α_i are zero, then

$$\sigma = \frac{1}{L}, \quad L = \inf_{\lambda} \left\| \sum_{i=1}^{n} \lambda_i x_i \right\|, \tag{5.46}$$

where the infimum is taken over all numbers λ which satisfy the constraint

$$\sum_{i=1}^{n} \lambda_i \alpha_i = 1.$$

If the vectors x_i, $i=1,\ldots,n$, are linearly independent, then the infimum is achieved and is non-zero. Indeed, applying the Bunyakovskii-Schwarz inequality to (5.47), we obtain that

$$1 = \sum_{i=1}^{n} \lambda_i \alpha_i \leq |\lambda| \cdot |\alpha| ,$$

where

$$|\lambda| = \left(\sum_{i=1}^{n} \lambda_i^2 \right)^{1/2}, \quad |\alpha| = \left(\sum_{i=1}^{n} \alpha_i^2 \right)^{1/2}.$$

Thus, from (5.47) it follows that

$$|\lambda| \geq \frac{1}{|\alpha|} .$$

Now set

$$\rho = \min_{|\lambda| = 1} \left\| \sum_{i=1}^{n} \lambda_i x_i \right\| .$$

Since the vectors x_i are linearly independent, $\rho > 0$. Then, for any non-zero vector $\lambda = (\lambda_1, \ldots, \lambda_n)$,

$$\left\| \sum_{i=1}^{n} \frac{\lambda_i}{|\lambda|} x_i \right\| \geq \rho ,$$

or

$$\left\| \sum_{i=1}^{n} \lambda_i x_i \right\| \geq \rho |\lambda| .$$

It follows from the last inequality that the domain of λ for which

$$\left\| \sum_{i=1}^{n} \lambda_i x_i \right\| \leq \left\| \sum_{i=1}^{n} \lambda_i^0 x_i \right\| \quad \text{and} \quad \sum_{i=1}^{n} \lambda_i^0 \alpha_i = 1$$

is contained in the domain

$$|\lambda| \leq \frac{\left\| \sum_{i=1}^{n} \lambda_i^0 x_i \right\|}{\rho} .$$

But it is clear that we must look for the minimum of $\| \sum_{i=1}^{n} \lambda_i x_i \|$ precisely in this domain. Since this domain is compact and the function to be minimized is continuous, the minimum is achieved. Moreover, this minimum is different from zero. Indeed, as we have already seen, if λ satisfies (5.47), then $|\lambda| \geq |\alpha|^{-1}$. Therefore, for all such λ,

$$\left\| \sum_{i=1}^{n} \lambda_i x_i \right\| \geq \rho |\lambda| \geq \rho |\alpha|^{-1} > 0.$$

The results which we have obtained permit us to establish a relation between the moment problem and the Chebyshev approximation problem.

Let there be given elements $\bar{x}, x_1, \ldots, x_n$ in a Banach space. Consider the moment problem

$$\min \|x^*\| ,$$
$$x^*(\bar{x}) = 1 ,$$
$$x^*(x_i) = 0 \text{ for } i=1,\ldots,n.$$

On the basis of the previous remark, the norm of a functional which is a solution to this problem equals

$$\sigma = \frac{1}{L} ,$$

where

$$L = \min_{\lambda_i} \left\| \bar{x} + \sum_{i=1}^{n} \lambda_i x_i \right\| , \quad i = 1, \ldots, n.$$

Indeed, in the problem under consideration, $\alpha_0 = 1$, $\alpha_i = 0$ for $i = 1, \ldots, n$, and, thus, Condition (5.47), which can be written in the form

$$\sum_{i=0}^{n} \lambda_i \alpha_i = 1,$$

is equivalent to the condition $\lambda_0 = 1$.

Therefore, finding a functional of minimum norm in the moment problem under consideration is equivalent to solving the problem of finding a best approximation to an element $\bar{x}$ by means of a linear combination $\sum_{i=1}^{n} \lambda_i x_i$.

Conversely, if the numbers λ_i are such that

$$\left\| \bar{x} + \sum_{i=1}^{n} \lambda_i x_i \right\| = L,$$

then, on the basis of Theorem 5.5 and its corollary, there exists a functional x_0^* such that

$$x_0^*\left(\bar{x} + \sum_{i=1}^{n} \lambda_i x_i\right) = \left\| \bar{x} + \sum_{i=1}^{n} \lambda_i x_i \right\| = L, \qquad (5.48)$$

with

$$\| x_0^* \| = 1 \text{ and } x_0^*(x_i) = 0 \text{ for } i=1,\ldots,n.$$

If $L \neq 0$ (and it is clear that only in this case is $\sigma < \infty$ and is the moment problem meaningful), then setting $x^* = L^{-1} x_0^*$, we obtain from (5.48) that

$$\|x^*\| = \sigma, \quad x^*(\bar{x}) = 1, \quad x^*(x_i) = 0,$$

i.e., x^* is a solution to the moment problem. Therefore, we have established that the moment problem and the Chebyshev approximation problem are the duals of one another, and that a solution to one of these problems can be obtained from a solution of the other.

Let us consider the special case of the moment problem in which the space under consideration is the space C of all continuous real-valued functions of an argument t which ranges over the interval [0,T]. In this space, the norm is given by the formula

$$\|x\| = \max_{0 \leq t \leq T} |x(t)| .$$

By a well-known theorem [2], the continuous functionals in C^* are of the form

$$x^*(x) = \int_0^T x(t)\, dg(t) ,$$

where g(t) is a function of bounded variation such that g(0) = 0, and

$$\|x^*\| = \operatorname*{Var}_{0 \leq t \leq T} g(t) .$$

In this connection, the moment problem can be formulated as follows: Find a function g(t) of minimum total variation on [0,T] which satisfies the conditions

$$\int_0^T \varphi_i(t)\, dg(t) = \alpha_i \quad \text{for } i=1,\ldots,n,$$

where the $\varphi_i(t)$ are given continuous functions, and the α_i are given numbers.

According to the remark to Theorem 5.10 the following problem corresponds to this problem:

$$\min_{\lambda} \max_{0 \leq t \leq T} \left| \sum_{i=1}^{n} \lambda_i \varphi_i(t) \right| ,$$

$$\sum_{i=1}^{n} \lambda_i \alpha_i = 1.$$

Thus, we are to find the minimum of the convex function

$$\mu(\lambda) = \max_{0 \leq t \leq T} \max_{-1 \leq \xi \leq +1} \xi \left(\sum_{i=1}^{n} \lambda_i \varphi_i(t) \right)$$

of a finite number of arguments on the set

$$\Omega = \left\{ \lambda : \sum_{i=1}^{n} \lambda_i \alpha_i = 1 \right\} .$$

We have already more than once come across functions such as $\mu(\lambda)$, e.g., in Sections 4 and 6 of this chapter. If we set

$$K(\lambda) = \left\{ t: 0 \leq t \leq T, \left| \sum_{i=1}^{n} \lambda_i \varphi_i(t) \right| = \mu(\lambda) \right\} ,$$

then the set of all support functionals to $\mu(\lambda)$ is the convex hull of all vectors of the form

$$\xi(t)\varphi(t) = (\xi(t)\varphi_1(t), \ldots, \xi(t)\varphi_n(t)),$$

where

$$\xi(t) = \operatorname{sign} \left(\sum_{i=1}^{n} \lambda_i \varphi_i(t) \right) \text{ for } t \in K(\lambda).$$

Further, it is easily seen that, at every point $\lambda \in \Omega$, the cone Γ_λ of admissible directions consists of all vectors e which satisfy the condition

$$\sum_{i=1}^{n} e_i \alpha_i = 0.$$

Therefore, the dual cone Γ_λ^* consists of all vectors of the form $\varkappa\alpha$, where

$$\alpha = (\alpha_1, \ldots, \alpha_n), \quad -\infty < \varkappa < +\infty .$$

Indeed, if $a \in \Gamma_\lambda^*$, then

$$(a, e) \geq 0$$

for every $e \in \Gamma_\lambda$, i.e., for every e which satisfies the condition

$$(e, a) = 0.$$

Let us represent a in the form $a = \varkappa\alpha + b$, where $(\alpha, b) = 0$. Such a representation is unique. Then

$$\varkappa(\alpha, e) + (b, e) \geq 0,$$

or,

$$(b, e) \geq 0$$

for every $e \in \Gamma_\lambda$. But $-b \in \Gamma_\lambda$, since $(-b, \alpha) = 0$. Thus,

$$-(b, b) = -\|b\|^2 \geq 0,$$

i.e., $b = 0$.

The function $\mu(\lambda)$ is continuous.

Suppose that the functions $\varphi_i(t)$ are linearly independent and that not all of the α_i are zero. Then, on the basis of the remark to Theorem 5.10, $\mu(\lambda)$ achieves its minimum on Ω, and this minimum is not zero.

Let $\lambda^0 \in \Omega$ be a minimum point. Then, on the basis of Theorem 2.1, there exists a functional in the set of support functionals to $\mu(\lambda)$ at x^0 which belongs to $\Gamma^*_{\lambda^0}$. It follows from the description of the sets of support functionals and of $\Gamma^*_{\lambda^0}$ that there exist numbers $\gamma_j \geq 0$, $j = 1, \ldots, \gamma$, a number $\varkappa$, and points $t_j \in K(\lambda^0)$ such that

$$\sum_{j=1}^{r} \gamma_j \xi(t_j)\varphi(t_j) = \varkappa\alpha, \quad \gamma_j \geq 0, \quad \sum_{j=1}^{r} \gamma_j = 1, \quad r \leq n+1. \quad (5.49)$$

Taking the inner product in (5.49) with λ^0, and taking into account the definitions of $K(\lambda)$ and $\xi(t)$ and the relation

$$(\lambda^0, \alpha) = 1,$$

we obtain

$$\varkappa = \mu(\lambda^0) = L > 0.$$

Now set

$$g_j(t) = \begin{cases} 0 & \text{for } 0 \le t < t_j, \\ 1 & \text{for } t_j \le t < T, \end{cases}$$

$$g(t) = \frac{1}{L} \sum_{j=1}^{r} \gamma_j \xi(t_j) g_j(t) .$$

Then, on the basis of the definition of the Stieltjes integral and (5.49), we have

$$\int_0^T \varphi(t)\, dg(t) = \frac{1}{L} \sum_{j=1}^{r} \gamma_j \xi(t_j) \varphi(t_j) = \alpha ,$$

or, in a "by-component" form,

$$\int_0^T \varphi_i(t)\, dg(t) = \alpha_i \quad \text{for } i = 1, \dots, n.$$

Further, $g(t)$ is a piecewise-constant function which has jumps at the points t_j of magnitude $\gamma_j \xi(t_j) L^{-1}$. Thus, the total variation of this function is simply equal to the sum of the absolute values of the jumps. But,

$$\sum_{j=1}^{r} | \gamma_j \xi(t_j) L^{-1} | = \frac{1}{L} \sum_{j=1}^{r} \gamma_j = \frac{1}{L} = \sigma.$$

Therefore, the norm of the functional which corresponds to $g(t)$ is equal to σ. Thus, we can state, on the basis of the remark to Theorem 5.10, that the just-constructed functional is a solution to the moment problem.

Theorem 5.11. *If n functions $\varphi_1(t), \ldots, \varphi_n(t)$ in C are linearly independent, then the moment problem always has a solution in the form of a function of bounded variation which is piecewise-constant and which has at most n+1 points of discontinuity.*

The moment problem may be applied in a number of ways to solve linear optimal control problems. In concluding this section, we shall briefly consider one such problem.

Let an object be described by the system of equations

$$\dot{x} = Ax + bu,$$

where A is an $n \times n$ matrix, b is an n-dimensional column vector, and u is a scalar-valued control function. We wish to transfer the system from an initial state x^0 to a final state x^1 in the time T, using an impulsive control with a minimum for the sum of the impulses. The use of an impulsive control means that u(t) has the form

$$u(t) = \sum_{j=1}^{k} \gamma_j \delta(t-t_j),$$

where the t_j are the times at which the impulses are applied, $|\gamma_j|$ is the "strength" of the j-th impulse and $\sum_{j=1}^{k} |\gamma_j|$ is the total "strength". The function $\delta(t)$ is the delta-function, which formally satisfies the following condition

$$\int_{\alpha}^{\beta} \varphi(t)\,\delta(t)\,dt = \varphi(0),$$

if $\alpha \leq 0$ and $\beta \geq 0$, and α and β are not both zero. If α and β have the same sign, then the integral vanishes. Obviously, if

$$g_0(t) = \begin{cases} 0 & \text{for } t < 0, \\ 1 & \text{for } t \geq 0, \end{cases}$$

then the value of the integral with the delta-function is quite the same as the value of the Stieltjes integral

$$\int_{\alpha}^{\beta} \varphi(t)\,dg_0(t).$$

Clearly, in connection with what we have just said, with every impulsive control we can associate a function of bounded variation of the form

$$g(t) = \sum_{j=1}^{k} \gamma_j g_0(t-t_j),$$

where

$$\operatorname{Var} g(t) = \sum_{j=1}^{k} |\gamma_j|.$$

Let us return to the originally stated problem. Every solution of a system of linear differential equations can be represented by virtue of the Cauchy formula in the form

$$x(t) = \Phi(t)x^0 + \int_0^t \Phi(t-\tau)bu(\tau)\,d\tau ,$$

where $d\Phi(t)/dt = A\Phi(t)$ and $\Phi(0)$ is the identity matrix. Thus, the given problem of transferring our system from x^0 to x^1 is equivalent to finding a control which satisfies the system of equations

$$x(T) = \Phi(T)x^0 + \int_0^T \Phi(T-\tau)bu(\tau)\,d\tau = x^1,$$

or

$$\int_0^T \varphi(\tau)u(\tau)\,d\tau = \alpha ,$$

where

$$\varphi(\tau) = \Phi(T-\tau)b, \text{ and}$$

$$\alpha = x^1 - \Phi(T)x^0 .$$

If we associate with $u(t)$ the function $g(t)$ according to the rules which we have just indicated, then the original problem is transformed into the following one: Find a function $g(t)$ of minimum total variation which satisfies the condition

$$\int_0^T \varphi(\tau)\,dg(\tau) = \alpha .$$

But this is precisely the moment problem in the space of continuous functions which we considered before. If $\alpha \neq 0$ (if $\alpha = 0$ then the solution is trivial: $u(t) \equiv 0$), and if the components of the vector function $\varphi(t)$ are linearly independent (which is also true if the "general position condition" holds, see [18, p. 116]), then, on the basis of Theorem 5.11, the moment problem which we have formulated has a piecewise-constant solution g(t) with at most n+1 points of discontinuity. But, to such a function there corresponds an impulsive control with at most n+1 impulses. The "strengths" of these impulses coincide with the magnitudes of the jumps of g(t), and the impulses are applied at the instants of time when g(t) is discontinuous. Thus, if $t_1, \ldots, t_n$ are the points of discontinuity of g(t), then

$$u^0(t) = \sum_{i=1}^{r} (g(t_j + 0) - g(t_j - 0))\, \delta(t - t_j), \quad r \leq n+1 .$$

The total strength of this control is equal to

$$\sum_{i=1}^{r} | g(t_j + 0) - g(t_j - 0) | = \operatorname{Var} g(t) .$$

Since the total variation of g(t) is minimal, $u^0(t)$ is an impulsive control with minimum total impulse.

Therefore, under our broad assumptions concerning the system of differential equations, an optimal control contains at most n+1 impulses. If we make use of results obtained in the course of proving Theorem 5.11, then we can assert that the instants of time at which the impulses are applied belong to the set

$$K(\lambda^0) = \{t: 0 \leq t \leq T, \ |\lambda^0, \varphi(t)| = L\},$$

where

$$L = \max_{0 \leq t \leq T} |\lambda^0, \varphi(t))| = \min_{(\lambda, \alpha) = 1} \max_{0 \leq t \leq T} |(\lambda, \varphi(t))|.$$

9. <u>A discrete maximum principle.</u> A broad class of control processes can be described in the following way: At discrete instants of time $t = 1, 2, \ldots,$ a control u_k is chosen such that, if the system was at the state x_k at the time k, then it goes over to the state x_{k+1} at the time k+1 as defined by the equation

$$x_{k+1} - f_k(x_k, u_k) = 0. \tag{5.50}$$

Here, for each k, x_k is an n-dimensional vector and $f_k(x, u)$ is an n-dimensional vector-valued function with components $f_k^i(x, u)$, $i = 1, \ldots, n$. The control u_k is chosen from a subset U of an r-dimensional space.

The optimization problem for the process which we have just described consists in choosing controls u_k, $k = 0, \ldots, N-1$, such that

$$\alpha_i(x_0) = 0 \quad \text{for } i = 1, \ldots, P_0 , \tag{5.51,1}$$

$$\varphi_k(x_k) \leq 0 \quad \text{for } k = 0, \ldots, N , \tag{5.51,2}$$

and such that the function $g_0(x_N)$ takes on its minimum value. With regard to the functions $\alpha_i(x_0)$ and $\varphi_k(x_k)$, we shall assume that they have continuous gradients which will be denoted by $\partial_{x_0} \alpha_i(x_0)$ and $\partial_{x_k} \varphi_k(x_k)$, respectively.

We now assume that the functions $f_k^i(x, u)$ are continuous in x and u and continuously differentiable with respect to x. Moreover, we shall assume that the sets $f_k(x, U) = \{ y: y = f_k(x, u),\ u \in U \}$ are convex for each x and k. Then it is easily seen that the problem which we have formulated is a special case of Problem (4.18). Indeed, if we introduce the vectors

$$x = (x_0, x_1, \ldots, x_N), \quad \text{and}$$
$$u = (u_0, u_1, \ldots, u_{N-1})$$

and the sets

$$X = \{ x: x_k \in E^n \text{ for } k = 0, \ldots, N \},$$

$$\bar{U} = \{u: u_k \in U \text{ for } k = 0, \ldots, N-1\},$$

then Equations (5.50) and (5.51,1) may be considered to be the equality-type constraints of Problem (4.18), and Inequalities (5.51,2) the inequality-type constraints. The function $g_0(x_N)$ corresponds to the function $\varphi_0(x, u)$ of Problem (4.18).

Now, let a vector e have the structure

$$e = (e_0, e_1, \ldots, e_N), \quad e_k \in E^n.$$

Then the directional differentials of the functionals appearing in the Constraints (5.50) and (5.51) in the direction e have the following forms, respectively:

$$\left.\begin{aligned} h_{f_k}(u, e) &= e_{k+1} - \partial_{x_k} f_k(x_k, u_k) e_k, \\ h_{\alpha_i}(u, e) &= \left(\partial_{x_0} \alpha_i(x_0), e_0\right), \\ h_{\varphi_k}(u, e) &= \left(\partial_{x_k} \varphi_k(x_k), e_k\right). \end{aligned}\right\} \qquad (5.52)$$

Here we have preserved vector notation for the Constraints (5.50), so that h_{f_k} is the vector with components

$$h_{f_k^i}(u, e) = e_{k+1}^i - \left(\partial_{x_k} f_k^i(x_k, u), e_k\right),$$

e_{k+1}^i is the i-th component of the vector e_{k+1}, $\partial_{x_k} f_k^i(x_k, u)$ is the gradient of the function $f_k^i(x_k, u)$ with respect to x_k, and the matrix $\partial_{x_k} f_k(x_k, u)$ has as its rows the gradients $\partial_{x_k} f_k^i(x_k, u)$.

Let us verify whether the basic Assumptions 1-4 of Theorem 4.6 are satisfied. In order that the first assumption be satisfied, it is necessary that, for each fixed x, the vectors

$$(g_0(x_N), x_1 - f_0(x_0, u_0), \ldots, x_N - f_{N-1}(x_{N-1}, u_{N-1}),$$
$$\alpha_1(x_0), \ldots, \alpha_{p_0}(x_0), \varphi_0(x_0), \ldots, \varphi_N(x_N))$$

form a convex set when u ranges over U. But this assumption does hold since the vectors $u_0, \ldots, u_{N-1}$ range over U independently, and, at the same time, each set $f_k(x_k, U)$ $(k = 0, \ldots, N-1)$ is convex.

Further, since $X = E^{(N+1)n}$, we can set

$$K_x = E^{(N+1)n},$$

so that the second assumption also holds.

Finally, Assumptions 3 and 4 are also satisfied, since all of the functions appearing in (5.50) and (5.51) are continuously differentiable.

Theorem 4.6 also requires that the functionals $h_i(u, e)$ which correspond to the equality-type constraints must be linearly independent.

Let us clarify what requirements this imposes on our problem. To do this, we suppose that these vectors are linearly dependent, i.e., that there exist numbers a_i, $i = 1, \ldots, P_0$, and vectors ψ_k, $k = 0, 1, \ldots, N-1$, such that

$$\sum_{i=1}^{P_0} a_i h_{\alpha_i}(u, e) + \sum_{k=0}^{N-1} (\psi_k, h_{f_k}(u, e)) = 0 \qquad (5.53)$$

for every e. Substituting into here Expressions (5.52), we obtain

$$\sum_{i=1}^{P_0} a_i \left(\partial_{x_0} \alpha_i(x_0), e_0 \right) + \sum_{k=0}^{N-1} \left[(\psi_k, e_{k+1}) - (\psi_k, \partial_{x_k} f_k(x_k, u_k) e_k) \right] = 0.$$

A regrouping of the terms yields

$$\left(\sum_{i=1}^{P_0} a_i \partial_{x_0} \alpha_i(x_0) - \left(\partial_{x_0} f_0(x_0, u_0) \right)^* \psi_0, e_0 \right) + \sum_{k=1}^{N-1} \left(\psi_{k-1} - \left(\partial_{x_k} f_k(x_k, u_k) \right)^* \psi_k, e_k \right) + (\psi_{N-1}, e_N) = 0.$$

(An asterisk after a matrix denotes transposition).

Now taking into account that the vectors e_k, $k=0, \ldots, N$, vary independently, we obtain the system of equations

$$\sum_{i=1}^{P_0} a_i \partial_{x_0} \alpha_i(x_0) = \left(\partial_{x_0} f_0(x_0, u_0)\right)^* \psi_0,$$

$$\psi_{k-1} - \left(\partial_{x_k} f_k(x_k, u_k)\right)^* \psi_k = 0 \text{ for } k = 1, \ldots, N-1,$$

$$\psi_{N-1} = 0 .$$

But it follows immediately from this system that $\psi_k = 0$ for each $k = 0, 1, \ldots, N-1$, so that, as a result, we obtain the equation

$$\sum_{i=1}^{P_0} a_i \partial_{x_0} \alpha_i(x_0) = 0.$$

If we suppose that the vectors $\partial_{x_0} \alpha_i(x_0)$ are linearly independent, then the last equation holds only if $a_i = 0$. But it already follows from this that all of the functionals h_{α_i} and h_{f_k} are linearly independent, since we have convinced ourselves that Relations (5. 53) hold only when the a_i and ψ_k are all equal to zero.

Theorem 5. 12. Let a control system be described by Eqs. (5. 50), and let $\bar{u}_k$, $k = 0, 1, \ldots, N-1$, and $\bar{x}_k$, $k = 0, 1, \ldots, N$, be, respectively, an optimal control and an optimal trajectory which minimize $g_0(x_N)$ subject to the Constraints (5. 51). Moreover, let us suppose that the functions $f_k(x_k, u)$ are continuously differentiable with respect to x_k, and that the same is true for the functions $\alpha_i(x_0)$ and $\varphi_k(x_k)$.

Then, if the sets

$$f_k(x_k, U) = \{f_k(x_k, u): u \in U\}, \quad k = 0, \ldots, N-1$$

are convex for every fixed x_k, and if the vectors $\partial_{x_0} \alpha_i(x_0)$, $i = 1, \ldots, P_0$, are linearly independent, there exist numbers $\beta_0 \geq 0$, $\lambda_k \geq 0$ $(k = 0, 1, \ldots, N)$, and a_i $(i = 1, \ldots, P_0)$, and vectors ψ_k $(k = 0, 1, \ldots, N-1)$, not all zero, such that

a) $$\left(\partial_{x_0} f_0(\bar{x}_0, \bar{u}_0)\right)^* \psi_0 = \sum_{i=1}^{P_0} a_i \partial_{x_0} \alpha_i(\bar{x}_0) + \lambda_0 \partial_{x_0} \varphi_0(\bar{x}_0),$$

b) $$\psi_{k-1} - \left(\partial_{x_k} f_k(\bar{x}_k, \bar{u}_k)\right)^* \psi_k + \lambda_k \partial_{x_k} \varphi_k(\bar{x}_k) = 0$$

for $k = 1, \ldots, N-1$,

c) $$\psi_{N-1} + \lambda_N \partial_{x_N} \varphi_N(\bar{x}_N) + \beta_0 \partial_{x_N} g_0(\bar{x}_N) = 0,$$

d) $$(\psi_k, f_k(\bar{x}_k, \bar{u}_k)) = \max_{u \in U} (\psi_k, f_k(\bar{x}_k, u))$$

for $k = 0, 1, \ldots, N-1$,

e) $$\lambda_k \varphi_k(\bar{x}_k) = 0 \text{ for } k = 0, \ldots, N.$$

Proof. As was shown previously, the requirements

which are imposed by the theorem hypotheses permit us to apply Theorem 4.6 to this problem. On the basis of this theorem, we can assert that there exist numbers $\beta_0 \geq 0$, $\lambda_k \geq 0$, and a_i, and vectors ψ_k such that

$$\lambda_k \varphi_k(\overline{x}_k) = 0 \text{ for } k = 0, 1, \ldots, N, \tag{5.54}$$

and

$$\beta_0 h_{g_0}(\overline{u}, e) + \sum_{i=1}^{P_0} a_i h_{\alpha_i}(\overline{u}, e) + \sum_{k=0}^{N} \lambda_k h_{\varphi_k}(\overline{u}, e) + \sum_{k=0}^{N-1} (\psi_k, h_{f_k}(\overline{u}, e)) \geq 0 \tag{5.55}$$

for all $e \in K_x$, and that also

$$\beta_0 g_0(\overline{x}) + \sum_{i=1}^{P_0} a_i \alpha_i(\overline{x}_0) + \sum_{k=0}^{N} \lambda_k \varphi_k(\overline{x}_k) + \sum_{k=0}^{N-1} (\psi_k, x_{k+1} - f_k(\overline{x}_k, \overline{u}_k)) = \min_{u \in U} \Big[\beta_0 g_0(\overline{x}) + \sum_{k=0}^{N} \lambda_k \varphi_k(\overline{x}_k) + \sum_{i=1}^{P_0} a_i \alpha_i(\overline{x}_0) + \sum_{k=0}^{N-1} (\psi_k, x_{k+1} - f_k(\overline{x}_k, u_k)) \Big]. \tag{5.56}$$

It follows from (5.54) that conclusion e) of the theorem holds.

Further, (5.56) is obviously equivalent to the condition

$$\sum_{k=0}^{N-1} (\psi_k, -f_k(\overline{x}_k, \overline{u}_k)) = \min_{u \in U} \sum_{k=0}^{N-1} (\psi_k, -f_k(\overline{x}_k, u_k)) .$$

This implies that conclusion d) of the theorem holds, inasmuch as every u_k varies independently.

Further, substituting Expressions (5.52) and

$$h_{g_0}(\bar{u}, e) = \left(\partial_{x_N} g_0(\bar{x}_N), e_N\right)$$

into (5.55), we obtain

$$\beta_0\left(\partial_{x_N} g_0(\bar{x}_N), e_N\right) + \sum_{i=1}^{P_0} a_i\left(\partial_{x_0} \alpha_i(\bar{x}_0), e_0\right)$$

$$+ \sum_{k=0}^{N} \lambda_k\left(\partial_{x_k} \varphi_k(\bar{x}_k), e_k\right)$$

$$+ \sum_{k=0}^{N-1} \left[(\psi_k, e_{k+1}) - (\psi_k, \partial_{x_k} f_k(\bar{x}_k, \bar{u}_k) e_k)\right] \geq 0.$$

A regrouping of the terms reduces this inequality to the form

$$\left(\sum_{i=1}^{P_0} a_i \partial_{x_0} \alpha_i(\bar{x}_0) + \lambda_0 \partial_{x_0} \varphi_0(\bar{x}_0) - (\partial_{x_0} f_0(\bar{x}_0, \bar{u}_0))^* \psi_0, e_0 \right)$$

$$+ \sum_{k=1}^{N-1} (\psi_{k-1} - (\partial_{x_k} f_k(\bar{x}_k, \bar{u}_k))^* \psi_k + \lambda_k \partial_{x_k} \varphi_k(\bar{x}_k), e_k)$$

$$+ (\psi_{N-1} + \beta_0 \partial_{x_N} g_0(\bar{x}_N) + \lambda_N \partial_{x_N} (\bar{x}_N), e_N) \geq 0.$$

This inequality must hold for every $e \in K_x$. But since K_x is the entire space and vectors e_k vary independently, the inequality can hold only if the coefficient of each e_k is zero. Equating the coefficients of e_0, e_k for $k = 1, \ldots, N-1$, and e_N to zero, we deduce that conclusions a), b) and c) of the theorem hold. |||

Condition d) of the theorem shows that, in discrete systems for which $f_k(x_k, U)$ is convex, a maximum principle in a form analogous to the maximum principle for systems described by ordinary differential equations [18] is satisfied. It is well known [47] that if $f_k(x_k, U)$ is not convex, then the statement which we have just made is not, generally speaking, true. For such systems, only a so-called local maximum principle can be obtained. This principle can be obtained with the aid of Theorem 4.1, if we assume that the functions $f_k(x_k, u)$ are differentiable with respect to u. We shall not dwell here on the derivation of this result, since it is much the same as the previous one. Instead, we shall present some arguments to show why a global maximum principle holds for systems which are described by ordinary differential equations.

Let a system be described by the system of ordinary differential equations

$$\frac{dx}{dt} - f(x, u) = 0, \qquad (5.57)$$

where x is an n-dimensional vector and f(x, u) is a function which is continuous in both of its arguments and is continuously differentiable with respect to x. The vector u ranges over a compact set U in an r-dimensional space. If we subdivide the time interval $[0, T]$ on which System (5.57) is to be considered into subintervals $[t_k, t_{k+1}]$ with $0 = t_0 < t_1 < \ldots < t_{N-1} < t_N = T$, then for each of these intervals we can write the equation

$$x(t_{k+1}) - x(t_k) - \int_{t_k}^{t_{k+1}} f(x(t), u(t))\, dt = 0,$$

which is obtained by integrating (5.57) over $[t_k, t_{k+1}]$. Now, if the subdivision is sufficiently fine, then, by the continuity of x(t) and f(x, u), the just-obtained equation can be approximated by the following one:

$$x_{k+1} - x_k - \int_{t_k}^{t_{k+1}} f(x_k, u(t))\, dt = 0,$$

where $x_k = x(t_k)$ for each $k = 0, 1, \ldots, N-1$.

We shall now consider the problem of finding a measurable control u(t), $0 \leq t \leq T$, with $u(t) \in U$ for all t, which minimizes $g_0(x_N)$, subject to the Constraints (5.51). To do this, we shall prove the following preliminary result.

Let there be given a continuous, n-dimensional, vector-valued function $g(u)$ of the argument u, where $u \in U$. Let us consider the mapping $u(t) \rightarrow z$ of an arbitrary measurable function $u(t)$ taking on its values in U into the n-dimensional space E^n. Let

$$Z = \left\{ z: z = \int_0^1 g(u(t))\, dt \right\}.$$

The set Z is convex. Indeed, let

$$z_1 = \int_0^1 g(u_1(t))\, dt ,$$

$$z_2 = \int_0^1 g(u_2(t))\, dt .$$

Let us show that $\lambda_1 z_1 + \lambda_2 z_2$ belongs to Z whenever λ_1 and $\lambda_2 > 0$ and $\lambda_1 + \lambda_2 = 1$, i.e., that there exists a function $u(t)$ such that

$$\lambda_1 z_1 + \lambda_2 z_2 = \int_0^1 g(u(t))\, dt.$$

To do this, we set

$$u(t) = \begin{cases} u_1\left(\frac{t}{\lambda_1}\right) & \text{for } 0 \le t < \lambda_1, \\ u_2\left(\frac{t - \lambda_1}{\lambda_2}\right) & \text{for } \lambda_1 < t \le 1. \end{cases}$$

Then,

$$\int_0^1 g(u(t))\,dt = \int_0^{\lambda_1} g\left(u_1\left(\frac{t}{\lambda_1}\right)\right) dt + \int_{\lambda_1}^1 g\left(u_2\left(\frac{t - \lambda_1}{\lambda_2}\right)\right) dt$$

$$= \lambda_1 \int_0^1 g(u_1(\tau))\,d\tau + \lambda_2 \int_0^1 g(u_2(\tau))\,d\tau = \lambda_1 z_1 + \lambda_2 z_2 ,$$

where we have made the following changes of variables in the first and second integrals, respectively:

$$\tau = \frac{t}{\lambda_1} ,$$

$$\tau = \frac{t - \lambda_1}{\lambda_2} .$$

It follows from the just-obtained result that, for fixed x_{k+1} and x_k, the left-hand side of Equation (5.58) ranges over a convex set as u(t) ranges over the class of measurable functions such that $u(t) \in U$ for every t. Now, if we suppose that the vectors $\partial_{x_0} \alpha_i(x_0)$, $i = 1, \ldots, P_0$, are linearly independent, then, just as was done when we considered

System (5.50), we can apply Theorem 4.6.

Here we shall not present any of the tedious manipulations which are entirely analogous to those made in the proof of Theorem 5.12. We shall only state the final result.

In order that a control u(t) and a trajectory $\bar{x}_k$ be optimal for System (5.58), subject to the Constraints (5.51), it is necessary that there exist numbers $\beta_0 \geq 0$, $\lambda_k \geq 0$, and a_i, and vectors ψ_k, not all zero, such that

a) $$\psi_0 + \left(\int_{t_0}^{t_1} \partial_x f(\bar{x}_0, \bar{u}(t))\,dt\right)^* \psi_0 = \sum_{i=1}^{P_0} a_i \partial_{x_0} \alpha_i(\bar{x}_0) + \lambda_0 \partial_{x_0} \varphi_0(\bar{x}_0),$$

b) $$\psi_{k-1} - \psi_k - \left(\int_{t_k}^{t_{k+1}} \partial_x f(\bar{x}_k, \bar{u}(t))\,dt\right) \psi_k + \lambda_k \partial_{x_k} \varphi_k(\bar{x}_k) = 0$$

for $k = 1, \ldots, N-1$,

c) $$\psi_{N-1} + \lambda_N \partial_{x_N} \varphi_N(\bar{x}_N) + \beta_0 \partial_{x_N} g_0(\bar{x}_N) = 0,$$

d) $$\left(\psi_k, \int_{t_k}^{t_{k+1}} f(\bar{x}_k, \bar{u}(t))dt\right) = \max_{u(t)\in U} \left(\psi_k, \int_{t_k}^{t_{k+1}} f(\bar{x}'_k, u(t))dt\right)$$

for $k = 0, 1, \ldots, N-1$,

e) $\lambda_k \varphi_k(\bar{x}_k) = 0$ for $k = 0, 1, \ldots, N$.

It is not difficult to see that Condition d) is equivalent to the following one:

$$(\psi_k, f(\bar{x}_k, \bar{u}(t)) = \max_{\nu \in U} (\psi_k, f(\bar{x}_k, \nu)) \tag{5.59}$$

almost everywhere on the interval $t_k \leq t \leq t_{k+1}$. Indeed, suppose that the following inequality holds on a set Γ of a positive Lebesgue measure:

$$(\psi_k, f(\bar{x}_k, \bar{u}(t))) < (\psi_k, f(\bar{x}_k, \nu_0)),$$

where $\nu_0 \in U$ is such that

$$(\psi_k, f(\bar{x}_k, \nu_0)) = \max_{\nu \in U} (\psi_k, f(\bar{x}_k, \nu)).$$

We choose a control u(t) of the form

$$u(t) = \begin{cases} \bar{u}(t) & \text{for } t_k \leq t \leq t_{k+1} \text{ and } t \notin \Gamma, \\ \nu_0 & \text{for } t \in \Gamma. \end{cases}$$

Then

$$\left(\psi_k, \int_{t_k}^{t_{k+1}} f(\bar{x}_k, \bar{u}_k(t))dt\right) = \int_{t_k}^{t_{k+1}} (\psi_k, f(\bar{x}_k, \bar{u}(t)))\,dt$$

$$= \int_{[t_k, t_{k+1}]\setminus\Gamma} (\psi_k, f(\bar{x}_k, \bar{u}(t)))dt + \int_{\Gamma} (\psi_k, f(\bar{x}_k, \bar{u}(t)))dt$$

$$< \int_{t_k}^{t_{k+1}} (\psi_k, f(\bar{x}_k, u(t)))dt,$$

which contradicts Condition d).

Relation (5.59) expresses a maximum principle for an optimal control of System (5.58). This principle has now been proved without any assumptions concerning the convexity of the set $f(x, U)$. The convexity of the left-hand side of (5.58) with respect to the controls $u(t)$ followed from the fact that an integration was performed.

Let us note that, under the condition

$$\Delta = \max |t_{k+1} - t_k| \to 0$$

the result which we have presented tends formally to the usual maximum principle [18].

SHORT BIBLIOGRAPHY

The bibliography which we shall present by no means pretends to contain the entire mass of literature that has been devoted to the theory of necessary conditions in extremal problems. In this bibliography, we have picked out only those works which pertain directly to the material which we presented or which, for some reasons or other, seemed to the author to be important to the development of the theory.

<u>For the introduction.</u> At present, there is a number of excellent and easily understood textbooks on Functional Analysis. The presentation of some of the basic facts of Functional Analysis in this book was based on the monograph of Dunford and Schwartz [1]. All these facts can also be found in the handbook [2]. A good presentation of Functional Analysis, as applied to the derivation of necessary conditions for an extremum, was given in the introduction to the article by L. Hurwicz [3]. A detailed theory of differentiation of operators and of functionals may be found in [7] and [8].

All of the basic information on convex sets in function

spaces is contained in Chapter V of the book [1]. An excellent presentation of the theory of convex cones and a proof of a separation theorem may be found in the article by M. G. Krein and M. A. Rutman [4]. The notion of a regularly convex set was first introduced by M. G. Krein and V. L. Šmulian [52].

The theory of convex sets in finite-dimensional spaces is presented in the book by S. Karlin [5]. A proof of the theorem on the convex hulls of sets in finite-dimensional spaces is also given there.

The properties of convex functions which were presented in the introduction can be found in the monograph of M. A. Krasnosel'skii and Ya. B. Rutickii [6].

For Chapter I. This chapter was written entirely on the basis of the article by B. N. Pschenichnyi [9], where sets of support functionals were introduced and their properties were thoroughly studied. In this paper, the relation between such sets and directional differentials was also established.

Similar sets were introduced for linearly convex functionals in the article by A. A. Milyutin and A. Y. Dubovitskii [10]. The relation between sets of support functionals and the theory of conjugate convex sets, which was developed by Fenchel (see [5] and [53]), should be noted. A number of properties of sets of support functionals, which are not contained in Chapter I, are presented in the work by E. G. Gol'shtein [14].

For Chapter II. Convex programming is the part of mathematical programming for which the theory is the most complete. The first important work in this field was the article by Kuhn and Tucker [54]. A proof of the Kuhn-Tucker theorem in most general form was given by H. Uzawa and L. Hurwicz [11] (see also [5]). The original proof of this theorem, due to Kuhn and Tucker, is presented in the book of J. B. Dennis [12]. This proof uses differential properties of the functions under consideration.

The proof of the Kuhn-Tucker theorem (Theorem 2. 5) presented in Chapter IV of this book is based on ideas of L. W. Neustadt [55]. This proof can also be obtained as an immediate consequence of Theorem 4. 6.

A differential form of this theorem was formulated in full extent by B. N. Pshenichnyi in [9] and [56].

Theorems 2. 1 - 2. 4 and 2. 6 may also be found in [9]. In connection with Theorem 2. 2, also see [10]. The corollary to Theorem 2.1, in the case where the convex functional is smooth, was established and exploited by V. F. Dem'yanov and A. M. Rubinov [13].

For Chapter III. The name "quasi-differentiable functionals" has been introduced here for the first time. The future will show to what extent this term will turn out to be successful. But it seemed necessary to in some way single out this broad class of functions which are met in extremal problems literally at every step.

The properties of functions obtained as the result of taking a maximum over a family of functions were studied in [15] and [56] in a simple case. These properties, as applied to convex functionals, were studied more completely in [9]. But, the proof which was given there did not, as a matter of fact, make use of convexity and could serve for a broader class of functionals.

Theorem 3.5 was proved in [9]. It can be generalized if the results of [14] are used.

For Chapter IV. As we indicated in the foreword, the main push in the development of the modern theory of necessary conditions for an extremum was given by optimal control theory. A maximum principle--a basic fact of this theory--was formulated by L. S. Pontryagin, V. G. Boltyanskii and R. V. Gamkrelidze in [16] and was proved in [17]. The monograph [18] is devoted to a complete presentation of the theory of optimal processes.

A number of original ideas related to proving the maximum principle may be found in the works of L. I. Rozenoer [19].

R. V. Gamkrelidze proved a maximum principle for systems with phase constraints under a number of assumptions concerning the structure of the problem in [20] and [21].

A general formulation of necessary conditions for an extremum in terms of conjugate cones was given for the first time by A. A. Milyutin and A. Y. Dubovitskii in [22]. The article [10] is devoted to an extensive presentation of their theory with applications to the theory of optimal processes. In [22] and [10], A. A. Milyutin and A. Y. Dubovitskii completely solved an optimal control problem under phase constraints.

In [23] and [57], R. V. Gamkrelidze introduced the notion of quasi-convex sets. This notion made it possible to give a simple and natural proof of the maximum principle.

Theorem 4.1 was proved by L. W. Neustadt and H. Halkin [58], under somewhat more general assumptions with regard to the set M. The proof which we presented in this book is different from that of [58].

A number of general theorems concerning necessary conditions for an extremum is contained in [55], [59], [60] and [61].

Theorem 4.6 was proved in [24].

<u>For Chapter V</u>. 1. Necessary conditions for an extremum in a finite-dimensional mathematical programming problem are presented in [12], and for convex programming in [25].

2. A minimization problem under a continuum of constraints was studied in [9] in the convex case.

3. Theorems concerning a minimax may be found in [5], where detailed references to original sources are given.

4. A large number of works is devoted to problems in Chebyshev approximation theory. A detailed bibliography for this problem may be found in the monograph of E. Y. Remez [26] and in the article by S. I. Zukhovitskii [27]. The basic ideas and methods of approach to the Chebyshev

approximation problem from the standpoint of Functional Analysis were worked out by M. G. Krein in [28]. The works [9], [14] and [29] are devoted to an investigation of various problems of a best approximation on the basis of a general theory of necessary conditions for an extremum.

5. A linear optimal control problem with phase constraints was studied in [30] and [56].

6. Dual problems of convex programming were studied by B. N. Pshenichnyi in [31] and [15] and by E. G. Gol'shtein in [32] and [33]. Another approach was used in [62], [63] and [12]. G. S. Rubinshtein [34] developed an original, general theory of duality.

7. A large literature is devoted to the theory of linear inequalities. The basic facts of this theory can be found in the works of S. N. Chernikov [35] and Ky Fan [36]. Systems of convex inequalities have been little studied. Helly's theorem, which is closely related to convex inequalities and which has numerous applications, is presented in the work of E. Helly [37]. A simple proof of this theorem was given by M. A. Krasnosel'skii [38].

8. The moment problem was systematically studied in the works of M. G. Krein [28]. He also established the dual

relationship between the moment problem and the Chebyshev approximation problem. This relationship was used by S. I. Zukhovitskii for the investigation of a problem of a best approximation.

N. N. Krasovskii demonstrated the relationship between the theory of moments and linear optimal control problems in [39], [40], and [41]. This relation was also used by R. Kulikowski [42], [43] and by P. Sarachik and G. Kranc [64].

Based on this relation, L. W. Neustadt obtained necessary conditions for optimality for the problem of optimal impulse control of a linear system, and investigated the structure of the optimal control [65].

9. A survey of the basic results concerning a discrete maximum principle, written by A. I. Propoi, is presented in Appendix 3 to the Russian translation of the book [48]. A local maximum principle for discrete systems was obtained by E. Polak and B. W. Jordan in [66] and by A. I. Propoi in [49]. H. Halkin [67] proved Theorem 5.12 under assumptions stronger than those made in Section 9 of Chapter V of this book. Halkin's approach was developed further in [68]. The relation between the discrete and continuous

maximum principles was investigated in [50].

Convexity (or quasi-convexity) properties of the images of sets obtained as the result of integral mappings were studied and made use of in extremal problems in [57], [69], [23], and [51].

LITERATURE

1. N. Dunford and J. T. Schwartz, Linear Operators. Part I: General Theory, Interscience, New York, 1958.

2. S. G. Krein (editor), Functional Analysis, Nauka, Moscow, 1964 (in Russian).

3. L. Hurwicz, Programming in linear spaces, in Studies in Linear and Non-linear Programming, by K. J. Arrow, L. Hurwicz and H. Uzawa, Stanford Univ. Press, Stanford, 1958, 38-102.

4. M. G. Krein and M. A. Rutman, Linear operators leaving invariant a cone in a Banach space, Uspekhi Mat. Nauk, t. 3 (1948), No. 1, 3-95 (English transl. in Amer. Math. Soc. Transl., 26 (1950)).

5. S. Karlin, Mathematical Methods and Theory in Games, Programming and Economics, Addison-Wesley, Reading, Mass., 1959.

6. M. A. Krasnosel'skii and Ya. B. Rutickii, Convex Functions and Orlicz Spaces, Noordhoff, Groningen, 1961.

7. L. A. Lusternik and V. I. Sobolev, Elements of Functional Analysis, F. Ungar, New York, 1961.

8. J. Dieudonné, Foundations of Modern Analysis, Academic Press, New York and London, 1960.

9. B. N. Pshenichnyi, Convex programming in a normed space, Kibernetika, No. 5 (1965), 46-54 (English transl. in Cybernetics, Vol. 1 (1965), No. 5, 46-57).

10. A. Y. Dubovitskii and A. A. Milyutin, Extremum pro-

blems in the presence of restrictions, Zh. Vychisl. Mat. i Mat. Fiz., t. 5 (1965), No. 3, 395-453 (English transl. in USSR Comput. Math. and Math. Phys., Vol. 5 (1965), No. 3, 1-80).

11. K. J. Arrow, L. Hurwicz and H. Uzawa, Studies in Linear and Non-linear Programming, Stanford Univ. Press, Stanford, 1958.

12. J. B. Dennis, Mathematical Programming and Electrical Networks, Technology Press, Cambridge, Mass., 1959.

13. V. F. Dem'yanov and A. M. Rubinov, Minimization of a smooth, convex functional on a convex set, Vestnik Leningrad. Univ., No. 19 (1964), 5-17.

14. E. G. Gol'shtein, Problems of best approximation by elements of a convex set and some properties of support functionals, Doklady Akad. Nauk SSSR, t. 173 (1967), No. 5, 995-999 (English transl. in Soviet Math. Dokl., Vol. 8 (1967), No. 2, 504-507).

15. B. N. Pshenichnyi, Dual method in extremal problems, I, Kibernetika, No. 3 (1965), 89-95 (English transl. in Cybernetics, Vol. 1 (1965), No. 3, 91-99).

16. V. G. Boltyanskii, R. V. Gamkrelidze and L. S. Pontryagin, On the theory of optimal processes, Doklady Akad. Nauk SSSR, t. 110 (1956), No. 1, 7-10.

17. V. G. Boltyanskii, The maximum principle in the theory of optimal processes, Doklady Akad. Nauk SSSR, t. 119 (1958), No. 6, 1070-1073.

18. L. S. Pontryagin, V. G. Boltyanskii, R. V. Gamkrelidze and E. F. Mishchenko, The Mathematical Theory of Optimal Processes, John Wiley, New York, 1962.

19. L. I. Rozenoer, L. S. Pontryagin's maximum principle in the theory of optimum systems I, II, III, Avtomat. i Telemekh., t. 20 (1959), No. 10, 1320-1334, No. 11, 1441-1458, No. 12, 1561-1578 (English transl. in Automat. Remote Control, Vol. 20 (1959), No. 10, 1288-1302,

No. 11, 1405-1421, No. 12, 1517-1532).

20. R. V. Gamkrelidze, Time-optimal processes with restricted phase coordinates, Doklady Akad. Nauk SSSR, t. 125 (1959), No. 3, 475-478.

21. R. V. Gamkrelidze, Time-optimal processes with restricted phase coordinates, Izv. Akad. Nauk SSSR Ser. Mat., t. 24 (1960), No. 3, 315-356.

22. A. Y. Dubovitskii and A. A. Milyutin, Extremum problems with constraints, Dokl. Akad. Nauk SSSR, t. 149 (1963), No. 4 (English transl. in Soviet Math. Dokl., Vol. 4 (1963), No. 2, 452-455).

23. R. V. Gamkrelidze, On the theory of the first variation, Dokl. Akad. Nauk SSSR, t. 161 (1965), No. 1, 345-348 (English transl. in Soviet Math. Dokl., Vol. 6 (1965), No. 2, 345-348).

24. B. N. Pshenichnyi, Necessary conditions for an extremum in partially convex programming problems, Kibernetika, No. 2 (1969), 90-93.

25. G. Zoutendijk, Methods of Feasible Directions, Elsevier, Amsterdam, 1960.

26. E. Y. Remez, General Computational Methods of Chebyshev Approximation, Publishing House of the Acad. of Sci. of the Ukrainian SSR, 1967.

27. S. T. Zukhovitskii, On the approximation of real functions in the sense of Chebyshev, Uspekhi Mat. Nauk, Vol. 11 (1956), No. 2, 125-159.

28. N. I. Ahiezer and M. Krein, Some Questions in the Theory of Moments, Amer. Math. Soc., Providence, 1962.

29. V. N. Tikhomirov, Some problems of approximation theory, Dokl. Akad. Nauk SSSR, t. 160 (1965), No. 4, 774-777 (English transl. in Sov. Math. Dokl., Vol. 6 (1965), No. 1, 202-205).

30. A. Y. Dubovitskii and A. A. Milyutin, Certain optimality problems for linear systems, Avtomat. i Telemekh., t. 24 (1963), No. 12, 1616-1626 (English transl. in Automat. Remote Control, Vol. 24 (1963), No. 12, 1471-1481).

31. B. N. Pshenichnyi, The duality principle in problems of convex programming, Zh. Vychisl. Mat. i Mat. Fiz., t. 5 (1965), No. 1, 98-106 (English transl. in USSR Comput. Math. and Math. Phys., Vol. 5 (1965), No. 1, 131-143).

32. E. G. Gol'shtein, Dual problems of convex and fractionally-convex programming in functional spaces, Doklady Akad. Nauk SSSR, t. 172 (1967), No. 5, 1007-1010 (English transl. in Soviet Math. Dokl., Vol. 8 (1967), No. 1,212-216).

33. E. G. Gol'shtein, Dual problems of convex programming, Ekonom. i Mat. Metody., No. 3 (1965), 410-425.

34. G.S. Rubinshtein, Dual extremum problems, Doklady Akad. Nauk SSSR, t. 152 (1963), No. 2, 288-291 (English transl. in Soviet Math. Dokl., Vol. 4 (1963), No. 5, 1309-1312).

35. S. N. Chernikov, Systems of linear inequalities, Uspekhi Mat. Nauk t. 8 (1953), No. 2, 7-73.

36. Ky Fan, On systems of linear inequalities, in Linear Inequalities and Related Systems, Kuhn and Tucker, eds., Annals of Mathematics Studies 38, Princeton Univ. Press, Princeton, 1956.

37. E. Helly, Über Mengen kuonvexer Körper mit gemeinschaftlichen Punkten, Jahrb. Deut. Math. Verein, Vol. 32 (1923), 175-176.

38. M. A. Krasnosel'skii, On a proof of Helly's theorem, Trudy Voronezhskogo Univ., t. 33 (1954).

39. N. N. Krasovskii, On the theory of optimum regulation, Avtomat. i Telemekh., t. 18 (1957), No. 11, 960-970 (English transl. in Automat. Remote Control, Vol. 18

(1957), No. 11, 1005-1016).

40. N. N. Krasovskii, A problem of optimal control, Prikl. Mat. Mekh., t. 21 (1957), No. 5, 670-677.

41. N. N. Krasovskii, On the theory of optimal control, Prikl. Mat. Mekh., t. 23 (1959), No. 4, 625-639 (English transl. in J. Appl. Math. Mech., Vol. 23 (1959), No. 4, 899-919).

42. R. Kulikowski, Optimizing processes and synthesis of optimizing automatic control systems with non-linear invariable elements, in Automatic and Remote Control, Proc. of the First Intern. Congress of the IFAC, Moscow, 1960, Vol. 1, 469-476, Butterworths, London, 1961.

43. R. Kulikowski, Optimal and Adaptive Processes in Automatic Control Systems, Nauka, Moscow 1967 (Russian transl. of Polish original).

44. G. M. Fikhtengol'tz, Course in Differential and Integral Calculus, t. 1, Fizmatgiz, 1959 (in Russian).

45. G. M. Fikhtengol'tz, Course in Differential and Integral Calculus, t. 2, Fizmatgiz, 1955 (in Russian).

46. S. Lefschetz, Differential Equations: Geometric Theory, Interscience, New York 1957.

47. A. G. Butkovskii, Distributed Control Systems, Academic Press, New York 1969.

48. L. T. Fan and C. S. Wang, The Discrete Maximum Principle: A Study of Multistage System Optimization, John Wiley, New York, 1964 (Russian transl. by Mir, 1967).

49. A. I. Propoi, The maximum principle for discrete control systems, Avtomat. i Telemekh., t. 26 (1965), No. 7, 1177-1187 (English transl. in Automat. Remote Control, Vol. 26 (1965), No. 7, 1167-1177).

50. R. Gabasov and F. M. Kirillova, Extending L. S.

Pontryagin's maximum principle to discrete systems, Avtomat. i Telemekh., No. 11 (1966), 46-51 (English transl. in Automat. Remote Control, Vol. 27 (1966), No. 11, 1878-1882).

51. V. I. Arkin, An infinite dimensional analog of non-convex programming problems, Kibernetika, No. 2 (1967), 87-93 (English transl. in Cybernetics, Vol. 3 (1967), No. 2, 70-75).

52. M. Krein and V. Šmulian, On regularly convex sets in the space conjugate to a Banach space, Ann. of Math., Vol. 41 (1940), 556-583.

53. W. Fenchel, Convex cones sets and functions, Office of Naval Research Logistics Project Report, Department of Math., Princeton University, 1953.

54. Proceedings of the Second Berkeley Symposium on Mathematical Statistics and Probability, J. Neyman, ed., University of California Press, Berkeley and Los Angeles, 1951, 481-492.

55. L. W. Neustadt, An abstract variational theory with applications to a broad class of optimization problems I: General theory, SIAM J. Control, Vol. 4 (1966), No. 3, 505-527.

56. B. N. Pshenichnyi, Linear optimal control problems, SIAM J. Control, Vol. 4 (1966), No. 4, 577-594.

57. R. V. Gamkrelidze, On some extremal problems in the theory of differential equations with applications to the theory of optimal control, SIAM J. Control, Vol. 3 (1965), No. 1, 106-128.

58. H. Halkin and L. W. Neustadt, General necessary conditions for optimization problems, Proc. Nat. Acad. Sci. USA, 56 (1966), 1066-1071.

59. H. Halkin, Nonlinear nonconvex programming in an infinite dimensional space, in Mathematical Theory of Control, A. V. Balakrishnan and L. W. Neustadt, eds.,

Academic Press, 1967, 10-25.

60. M. Canon, C. Cullum and E. Polak, Constrained minimization problems in finite dimensional spaces, SIAM J. Control, Vol. 4 (1966), 528-547.

61. O. L. Mangasarian and S. Fromovitz, A maximum principle in mathematical programming, in Mathematical Theory of Control, A. V. Balakrishnan and L. W. Neustadt, eds., Academic Press, 1967, 85-95.

62. O. L. Mangasarian, Duality in nonlinear programming, Quart. Appl. Math., Vol. 20 (1962), No. 3, 300-302.

63. P. Wolfe, A duality theorem for nonlinear programming, Quart. Appl. Math., Vol. 19 (1961), 239-244.

64. G. M. Kranc and P. E. Sarachik, An application of functional analysis to the optimal control problems, Trans. ASME Ser. D. J. Basic Eng., Vol. 85 (1963), No. 2, 143-150.

65. L. W. Neustadt, Optimization, a moment problem and nonlinear programming, SIAM J. Control, Vol. 2 (1964), No. 1, 33-53.

66. B. W. Jordan and E. Polak, Theory of a class of discrete optimal control systems, J. Electronics Control, Vol. 17 (1964), No. 6, 697-711.

67. H. Halkin, A maximum principle of the Pontryagin type for systems described by nonlinear difference equations, SIAM J. Control, Vol. 4 (1966), No. 1, 90-112.

68. J. Holtzman and H. Halkin, Directional convexity and the maximum principle for discrete systems, SIAM J. Control, Vol. 4 (1966), No. 2, 263-275.

69. J. Warga, Relaxed variational problems, J. Math. Anal. Appl., Vol. 4 (1962), No. 1, 111-128.

NOTES AND SUPPLEMENTARY BIBLIOGRAPHY TO AMERICAN EDITION

The main text of the book was written in 1967. Since then, a number of new articles devoted to the theory of necessary conditions for an extremum has appeared. Moreover, the author has become acquainted with a number of works on the theory of convex functions of which he was unaware at the time he wrote the book. In what follows, some notes which, I hope, will help the reader to establish the relation between the content of this book and some new results, will be presented.

All references in the remainder of the text pertain to the supplementary list of references.

For Chapter I. In this book, we only considered bounded convex functionals which are defined on the entire space. A theory of convex functions which are also allowed to take on the value $+\infty$ has been constructed in the works of Moreau [13] and Rockafellar [17-23].

A survey of the results of the theory of such functions

may be found in [13] and [17], and in the article by Ioffe and Tikhomirov [5]. The broadening of the class of convex functions under consideration makes it possible to obtain a number of results which are more general than those obtained in this book. It also permits one to only consider convex problems without constraints, since constraints can be taken into account by the fact that the function being minimized takes on the value $+\infty$ outside of the admissible region.

Thus, in particular, a result more general than Theorem 1.3 was obtained in [18], [20], and [23]. It was shown in [23] how Theorem 1.4 follows from the generalized Theorem 1.3. Some analytic properties of sets of support functionals ("subgradients" in the terminology of Rockafellar) were studied in [22].

Sets of support functionals to functions which arise in problems of the Calculus of Variations were studied in [6] and [21].

For Chapter II. A result analogous to Theorem 2.3 is found in the work of Rockafellar [23]. This result was obtained with the use of a generalization of Theorem 1.3. A special case of Theorem 2.4 was presented in [20].

For Chapter III. Theorem 3.2 was generalized in various directions by Sotskov in [10]. The results of Theorems 3.2 and 3.5 in the case of a family of convex functions were substantially developed by Levin in [7]. Particularly, he showed that the result of Theorem 3.5 (if $\varphi(x, \alpha)$ is a function which is convex in x for every α) holds also without the assumption of Gâteaux differentiability, and that the requirement of the uniform convergence of $\gamma(\lambda, \alpha)$ to zero in Theorem 3.5 can be deleted in the convex case. As is shown by some examples, this is impossible for functions $\varphi(x, \alpha)$ which are not convex in x.

For Chapter IV. Gamkrelidze, in [11] and [12], obtained some necessary conditions for an extremum using topological methods. Neustadt in [15] applied the theory which he had constructed (see the article [55] of the main list of literature) to optimal control problems. In the articles [3] and [4], Dubovitskii and Milyutin investigated the optimal control problem in the case where there are mixed constraints on phase coordinates and control of the form

$$g(x, u) \leq 0,$$

or

$$g(x, u) = 0.$$

In Neustadt's article [16], the general theory of necessary conditions was built on the basis of the notion of (φ, ϕ, Z)-extremality, introduced by himself. It is shown that the problem of obtaining necessary conditions is equivalent to obtaining conditions for (φ, ϕ, Z)-extremality.

In this book, basically (but not exclusively; see Theorems 4.4 and 4.5) we considered the case where the constraints are given in the form of a system of equalities for some functionals. In [16] and [24], there were studied problems which contained constraints of the form

$$A(x) \leq 0,$$

where $A: B \to B_1$ is a non-linear operator from a Banach space into another, and the relation $\geq$ in B_1 is induced by some closed, convex cone K_1: $y \geq 0$ if and only if $y \in K_1$. If we set

$$M^* = \overline{co} \{ y^*: y^* \in K_1^*, \; \|y^*\| = 1 \},$$

and

$$\mu(y) = \max_{y^* \in M^*} y^*(y),$$

then the relation $A(x) \leq 0$ is equivalent to the inequality

$\mu(A(x)) \leq 0$.

Therefore, an operator constraint can be given by an inequality for some functional. If K_1 contains interior points, then $0 \notin M^*$, and, making use of the inequality $\mu(A(x)) \leq 0$ together with Theorems 1.6 and 3.1, we are led to effective conditions for an extremum for the problem with operational constraints. If K_1 does not contain interior points, then a number of fundamental difficulties arises. Levin [8] succeeded in overcoming such difficulties only in the case where the operator $A(x)$ and the functional being minimized are linear.

Necessary conditions for an extremum in a finite-dimensional space were presented in various forms in the book by Mangasarian [14]. Here, great attention was paid to the description of conditions under which the coefficient λ_0 in Theorem 4.1 does not vanish.

Finally, we note that the requirement of linear independence of the functionals $h_i(u_0, e)$ in Theorem 4.6 can be deleted. However, it is true that, in order to do this, the entire proof must be significantly rebuilt.

<u>For Section 4 of Chapter V.</u> Some new applications of necessary conditions for an extremum to problems in

approximation theory may be found in [2].

<u>For Section 6 of Chapter V.</u> A duality theory in convex programming problems was systematically studied by Rockafellar in [18], [19], and [20]. It seems that the most general formulation of the duality principle is due to Gol'shtein [1]. A duality theory for linear programming problems in an infinite dimensional space was developed by Van Slyke and Wets [25].

<u>For Section 9 of Chapter V.</u> According to the notes for Chapter 4, the hypotheses on the linear independence of the vectors $\partial_{x_0} \alpha_i(x_0)$ in Theorem 5.12 can be deleted.

SUPPLEMENTARY LITERATURE

1. E. G. Gol'shtein, Dual problems of convex and fractionally convex programming in functional spaces, in the collection of papers, Investigations in Mathematical Programming, Nauka, Moscow 1968 (in Russian).

2. E. G. Gol'shtein, On problems of the best approximation by elements of a convex set, in the collection of papers, Investigations in Mathematical Programming, Nauka, Moscow 1968 (in Russian).

3. A. Y. Dubovitskii and A. A. Milyutin, Necessary conditions for a weak minimum in optimal control problems with mixed inequality-type constraints, Zh. Vychisl.Mat. i Mat. Fiz., t. 8 (1968), No. 4, 725-779.

4. A. Y. Dubovitskii and A. A. Milyutin, Maximum principle in a class of variations of small absolute value for optimal control problems with mixed constraints of equality and inequality, Doklady Akad. Nauk SSSR, t. 189 (1969), No. 6, 1177-1180 (English transl. in Soviet Math. Doklady, Vol. 10 (1969), No. 6, 1567-1571).

5. A. D. Ioffe and V. M. Tikhomirov, Duality of convex functions and extremum problems, Uspekhi Mat. Nauk, t. 23 (1968), No. 6, 51-116 (English transl. in Russian Math. Surveys, Vol. 23 (1968), No. 6, 53-124).

6. A. D. Ioffe and V. M. Tikhomirov, On minimization of integral functionals, Funkcional. Anal. i Prilozhen., t. 3 (1969), No. 3 (English transl. in Functional Anal. Appl., Vol. 3 (1969), No. 3, 218-227).

7. V. L. Levin, Some properties of support functionals, Mat. Zametki, t. 4 (1968), No. 6 (English transl. in Math.

Notes, Vol. 4 (1968), No. 5-6, 900-906).

8. V. L. Levin, Conditions for an extremum in infinite-dimensional linear problems with operator constraints, in the collection of papers, Investigations in Mathematical Programming, Nauka, Moscow 1968 (in Russian).

9. A. I. Sotskov, On the differentiability of a functional in a programming problem in an infinite-dimensional space, Kibernetika, No. 3 (1969), 81-88.

10. A. I. Sotskov, Necessary conditions for a minimum for a type of non-smooth problems, Doklady Akad. Nauk SSSR, t. 189 (1969), No. 2 (English transl. in Soviet Math. Doklady, Vol. 10 (1969), No. 6, 1410-1413).

11. R. V. Gamkrelidze and G. L. Kharatishvili, Extremal problems in linear topological spaces. I. Math. Systems Theory, Vol. 1 (1967), No. 3, 229-256.

12. R. V. Gamkrelidze, Extremal problems in finite-dimensional spaces, J. Optimization Theory Appl., Vol. 1 (1967), No. 3, 173-193.

13. J. J. Moreau, Fonctionelles convexes, mimeographed lecture notes, Collège de France, 1967.

14. O. L. Mangasarian, Nonlinear Programming, McGraw-Hill, New York, 1969.

15. L. W. Neustadt, An abstract variational theory with applications to a broad class of optimization problems, II. Applications, SIAM J. Control, Vol. 5 (1967), No. 1, 90-137.

16. L. W. Neustadt, A general theory of extremals, J. Comput. System Sci., Vol. 3 (1969), No. 1, 57-92.

17. R. T. Rockafellar, Convex Analysis, Princeton Univ. Press, 1970.

18. R. T. Rockafellar, Convex functionals and duality in optimization problems and dynamics, Lecture Notes in

Operations Research and Mathematical Economics, Vol. 11, Springer-Verlag, Berlin-Heidelberg - New York, 1969, 401-422.

19. R. T. Rockafellar, Duality and stability in extremum problems involving convex functions, Pacific J. Math., Vol. 21 (1967), No. 1, 167-187.

20. R. T. Rockafellar, An extension of Fenchel's duality theorem for convex functions, Duke Math. J., Vol. 33 (1966), No. 1, 81-90.

21. R. T. Rockafellar, Integrals which are convex functionals, Pacific J. Math., Vol. 24 (1968), No. 3, 867-873.

22. R. T. Rockafellar and A. Brønsted, On the subdifferentiability of convex functions, Proc. Amer. Math. Soc., Vol. 16 (1965), No. 4, 605-611.

23. R. T. Rockafellar, Convex functions, monotone operators and variational inequalities, Theory and Applications of Monotone Operators, Proceedings of NATO Advanced Study Institute held in Venice, Italy, June 17-30, 1968.

24. Y. Sakawa and Y. Nagahisa, Nonlinear programming in Banach spaces, J. Optimization Theory Appl., Vol. 4 (1969), No. 3, 182-190.

25. R. M. Van Slyke and R. J.-B. Wets, A duality theory for abstract mathematical programs with applications to optimal control theory, J. Math. Anal. Appl., Vol. 22 (1968), No. 3, 679-706.

SUBJECT INDEX

A

Adjoint operator, 21
Admissible controls, 142

B

Banach space, 17
Basis of the strong topology, 12
Basis of the topology, 2
Basis of the weak* topology, 14
Bounded functionals, 39
Bounded set, 13

C

Cauchy formula, 145
Chebyshev approximation theory, classical result of, 138
  fundamental theorem, 139
  in a Banach space, 190
Closed convex hull of a set, 26
Closed set, 3
Closure, 3
  in the strong topology, 15
  in the weak* topology, 15
Compactness, 6
  in the strong topology, 15
  in the weak* topology, 15
  sequential, 6
Complete space, 17
Complete normed space, 17
Cone, convex, 30
  conjugate, 30
  dual, 30

properties of, 31
Conjugate space, 11
Continuous mapping, 5
Convergence of sequences, 3
strong, 14
weak*, 14
Convex combination of points, 27
Convex cone Γ_{x_0}, 54
Convex functional, 35
Convex hull of a set, 26
Convex programming problems, 54
Convex set, 22
Cover of a set, 7

D

Differential of an operator, 20
Fréchet, 21
Gâteaux, 21
directional, 38, 42
Direct product of two spaces, 7
Directional differential of a functional, 38, 42
Dubovitskii and Milyutin, theory of, 95
first theorem of, 96
second theorem of, 98
Dual cone, 30

F

Fréchet differential, 21
Functionals, 11
bounded, 39
continuous linear on E^n, 18
continuous linear on C, 20
linear, 12
convex, 35
quasi-differentiable, 68
strong convergence of, 14
support, 39
weak* convergence of, 14
Fundamental sequence, 16

G

Gâteaux differential, 21

H

Hausdorff topological space, 3
Helly's theorem, 166

I

Image, 4
Interior point, 2

K

Kuhn-Tucker theorem, 65, 93

L

Lagrange multiplier rule, 82
Linear space, 8
Linear topological space, 9

M

Mapping, 4
  continuous, 5
  continuous at a point, 5
Mathematical programming problems, 82
Metric space, 6
Minimax problems, 126
Minimum of a convex bounded functional, 56
  on the entire space, 62
  under inequality constraints, 63
Multiplication of a set by a scalar, 17

N

Neighborhood of a point, 2
  of a set, 2
Norm, 10
Normed space, 10
  complete, 17

O

Open set, 2
Operator, 11
  adjoint, 21
  differential of, 20
  Fréchet differentiable, 21
  Gâteaux differentiable, 21
  linear, 21

P

Pre-image, 4
Product, direct of two spaces, 7
Programming, convex, 59
  mathematical, 82

Q

Quasi-differentiable functionals, 68

R

Regularly convex set, 25

S

Scalar product, 19
Sequences, convergence of, 3
Sequential compactness, 6
Separation theorem, 24
Set, open, 2
  bounded, 13
  closed, 3
    in the strong topology, 15
    in the weak* topology, 15
  compact, 7
  convex, 22
  regularly convex, 25
Solvability of a system of non-linear equations, 114
Space, topological, 2
  Banach, 17
  complete, 17

complete normed, 17
conjugate, 11
Hausdorff topological, 3
linear, 8
linear topological, 9
metric, 6
normed, 10
Strong convergence of a sequence of functionals, 14
Strong topology, 12
basis of, 12
Subcover, finite, 7
Sum of two subsets in a linear space, 17
Support functionals, 39

T

Topological space, 2
Hausdorff, 3
Topology, 2
basis of, 2
strong, 12
basis of, 12
weak*, 13
basis of, 14

V

Vandermonde determinant, 137

W

Weak* convergence, 14
Weak* topology, 13
basis of, 14
Weierstrass theorem in analysis, 7